Frontiers in Mental Health and the Environment

Frontiers in Mental Health and the Environment

Special Issue Editor

Marco Helbich

MDPI • Basel • Beijing • Wuhan • Barcelona • Belgrade

MDPI

Special Issue Editor
Marco Helbich
Utrecht University
The Netherlands

Editorial Office
MDPI
St. Alban-Anlage 66
4052 Basel, Switzerland

This is a reprint of articles from the Special Issue published online in the open access journal *International Journal of Environmental Research and Public Health* (ISSN 1660-4601) from 2017 to 2018 (available at: https://www.mdpi.com/journal/ijerph/special_issues/mental_health)

For citation purposes, cite each article independently as indicated on the article page online and as indicated below:

LastName, A.A.; LastName, B.B.; LastName, C.C. Article Title. *Journal Name* **Year**, *Article Number, Page Range.*

ISBN 978-3-03897-390-4 (Pbk)
ISBN 978-3-03897-391-1 (PDF)

Cover image courtesy of istock.com user Daniel V. Fung.

Contents

About the Special Issue Editor . . . . **vii**

Marco Helbich
Mental Health and Environmental Exposures: An Editorial
Reprinted from: *International Journal of Environmental Research and Public Health* **2018**, *15*, 2207,
doi:10.3390/ijerph15102207 . . . . **1**

Vikram Nichani, Kim Dirks, Bruce Burns, Amy Bird and Cameron Grant
Green Space and Depression during Pregnancy: Results from the *Growing Up in New Zealand*
Study
Reprinted from: *International Journal of Environmental Research and Public Health* **2017**, *140*, 1083,
doi:10.3390/ijerph14091083 . . . . **5**

Susanne Boers, Karin Hagoort, Floortje Scheepers and Marco Helbich
Does Residential Green and Blue Space Promote Recovery in Psychotic Disorders?
A Cross-Sectional Study in the Province of Utrecht, The Netherlands
Reprinted from: *International Journal of Environmental Research and Public Health* **2018**, *15*, 2195,
doi:10.3390/ijerph15102195 . . . . **23**

Rebecca M. Schwartz, Christina N. Gillezeau, Bian Liu, Wil Lieberman-Cribbin and Emanuela Taioli
Longitudinal Impact of Hurricane Sandy Exposure on Mental Health Symptoms
Reprinted from: *International Journal of Environmental Research and Public Health* **2017**, *14*, 957,
doi:10.3390/ijerph14090957 . . . . **31**

Rebecca M. Schwartz, Stephanie Tuminello, Samantha M. Kerath, Janelle Rios, Wil Lieberman-Cribbin and Emanuela Taioli
Preliminary Assessment of Hurricane Harvey Exposures and Mental Health Impact
Reprinted from: *International Journal of Environmental Research and Public Health* **2018**, *15*, 974,
doi:10.3390/ijerph15050974 . . . . **43**

Julián Alfredo Fernández-Niño, Víctor Alfonso Flórez-García, Claudia Iveth Astudillo-García and Laura Andrea Rodríguez-Villamizar
Weather and Suicide: A Decade Analysis in the Five Largest Capital Cities of Colombia
Reprinted from: *International Journal of Environmental Research and Public Health* **2018**, *15*, 1313,
doi:10.3390/ijerph15071313 . . . . **52**

Hoehun Ha and Wei Tu
An Ecological Study on the Spatially Varying Relationship between County-Level Suicide Rates
and Altitude in the United States
Reprinted from: *International Journal of Environmental Research and Public Health* **2018**, *15*, 671,
doi:10.3390/ijerph15040671 . . . . **63**

Ruoyu Wang, Desheng Xue, Ye Liu, Penghua Liu and Hongsheng Chen
The Relationship between Air Pollution and Depression in China: Is Neighbourhood Social
Capital Protective?
Reprinted from: *International Journal of Environmental Research and Public Health* **2018**, *15*, 1160,
doi:10.3390/ijerph15061160 . . . . **79**

Nena Kopčavar Guček and Polona Selič
Depression in Intimate Partner Violence Victims in Slovenia: A Crippling Pattern of Factors
Identified in Family Practice Attendees
Reprinted from: *International Journal of Environmental Research and Public Health* **2018**, *15*, 210,
doi:10.3390/ijerph15020210 . . . . **93**

Yang Xiao, Siyu Miao, Chinmoy Sarkar, Huizhi Geng and Yi Lu
Exploring the Impacts of Housing Condition on Migrants' Mental Health in Nanxiang,
Shanghai: A Structural Equation Modelling Approach
Reprinted from: *International Journal of Environmental Research and Public Health* **2018**, *15*, 225,
doi:10.3390/ijerph15020225 . . . . **109**

Darren J. Mayne, Geoffrey G. Morgan, Bin B. Jalaludin and Adrian E. Bauman
Does Walkability Contribute to Geographic Variation in Psychosocial Distress? A Spatial
Analysis of 91,142 Members of the 45 and Up Study in Sydney, Australia
Reprinted from: *International Journal of Environmental Research and Public Health* **2018**, *15*, 275,
doi:10.3390/ijerph15020275 . . . . **123**

About the Special Issue Editor

Marco Helbic, Dr., is an Associate Professor in the Department of Human Geography and Spatial Planning, Utrecht University. Prior to his appointment at Utrecht University, he was an Alexander von Humboldt research fellow at Heidelberg University, Germany, where he was also awarded with the venia legendi in human geography in 2015. Before this, he was a visiting scholar at the Louisiana State University, USA, and worked at the Austrian Academy of Sciences. In 2009, he obtained his Ph.D. (summa cum laude) at the University of Vienna. Dr. Helbich has developed a strong interest in urban mental health geographies, and combines his expertise in urban and health geography as well as geoinformatics. His recent research, funded by the European Research Council, addresses how environmental exposures may act as potential stressors or buffers to people's mental health, emphasizing people's daily mobility and their residential life-course.

International Journal of
*Environmental Research
and Public Health*

MDPI

Editorial

Mental Health and Environmental Exposures: An Editorial

Marco Helbich

Department of Human Geography and Spatial Planning, Faculty of Geosciences, Utrecht University, Princetonlaan 8a, 3584 CB Utrecht, The Netherlands; m.helbich@uu.nl; Tel.: +31-30-253-2017

Received: 26 September 2018; Accepted: 28 September 2018; Published: 10 October 2018

Keywords: health geography; mental disorders; exposures; risk assessment; environments; environmental modelling

1. Introduction

It is well-documented that human mental health emerges from a complex interplay between genetic, psychological, lifestyle, and other factors. In addition, people are also exposed to numerous environments. These environmental exposures (e.g., green space, noise, air pollution, weather conditions, housing conditions) might trigger mental disorders or be protective factors, facilitating stress reduction, mental recovery, etc. [1,2]. In this special issue, "environmental exposure" is understood in the broadest sense, comprising natural (e.g., park, bodies of water, weather) [3], social (e.g., capital, cohesion) [4], and built environmental exposures (e.g., urbanicity, intersection density, land use mix) [5]. Although some environmental factors—e.g., air pollution and green space—have already received broad attention in scientific debates, others have received very little, resulting in a tentative and partly inconclusive understanding of the environment–mental health relationship.

Mental illness contributes significantly to the global burden of mental disorders (i.e., 13% disability adjusted life-years lost) [6]. It is therefore important to grasp how and to what extent environmental exposures affect mental health outcomes. In the past year, 20% of all adults worldwide suffered from a mental disorder. Mental disorders have a lifetime prevalence of two out of seven adults and will continue to remain a leading cause of disease burden [7]. Such disorders have devastating consequences for people's quality of life and represent striking challenges for health systems as a whole. Thus, the reduction of mental disorders is a health priority in both developed and developing countries.

The geographic context of individuals is a central construct in assessing the contribution of environmental exposures to people's mental health [2]. While residential neighborhoods are frequently thought to represent an environmental context, this approach is increasingly critiqued because it assumes that people are immobile and exposed only to their residential neighborhoods. As this seems to be too restrictive an assumption, mobility-based environmental exposure assessments in mental health research have been put forward as methods that represent exposures more accurately. Such approaches highlight the importance of exposures that people experience throughout the day and over their lifetime [2].

2. Objective of the Special Issue

The collection of international case studies presented in this special issue contribute to a better understanding of which environmental exposures affect mental health outcomes, as well as how and to what extent they do so. These case studies provide novel insights into the interaction between mental health and the environment (e.g., green space and natural disasters). To present state-of-the-art methods and to further stimulate lively discussions on this topic, scholars were invited to submit original research, methodological papers, reviews, and meta-analyses related to the entire spectrum of mental disorders (e.g., depression, schizophrenia). This special issue also features papers documenting how scientific findings are translated into prevention strategies, health policies, and clinical practices.

3. The Papers

By the time of the submission deadline (i.e., the end of August 2018), a total of 10 manuscripts were accepted after a single-blind review process by at least two international experts using the journal-specific review guidelines. As usual, the scientific quality of the research and its methodological soundness had a crucial influence on whether a manuscript was accepted. If major revisions were requested by the reviewers, or needed to guarantee high scientific quality, a second review of the revised manuscript was conducted by at least one of the original reviewers or an alternative reviewer. If a review called for only minor revisions, a second review was not conducted. Instead, the guest editor decided whether the revised manuscript was fit for publication.

The first study by Nichani and colleagues [8] used cohort data from New Zealand to investigate whether the distance of an individual's residential location to the nearest green space affects depression risk during pregnancy. No evidence was found to support the hypothesis that maternal exposure to green space lowers the risk of antenatal depression. Similarly, after investigating patients in Utrecht, Netherlands, Boers et al. [9] found no significant associations between hospital admissions for psychotic disorders and exposure to green and blue space. The experience of natural disasters such as hurricanes, however, can have long-lasting effects on people's mental health outcomes. A two-paper series by Schwartz and colleagues addressed this by studying the impact of hurricanes Sandy [10] and Harvey [11] on numerous mental health symptoms using New York City and Long Island residents' data. Longitudinal analysis provided evidence that, for example, personal and property damage caused by hurricanes evoked symptoms of post-traumatic stress disorders but, in the case of Hurricane Sandy, not anxiety and depression symptoms. Natural disasters are not the only occurrences to have adverse effects on mental health outcome. Daily weather conditions are increasingly reported to influence suicide mortality [12]. No evidence for associations between suicide risk, daily temperature, and rainfall was found by Fernández-Niño et al. [13] for Columbian cities. Scientific evidence is mounting that mental health, in general, and suicide mortality, in particular, are related not only to personal characteristics and life events but also to environmental exposures other than weather conditions [14,15]. Two examples are reported in this special issue. Firstly, in a nationwide ecological study of the USA, Ha and Tu [16] showed that altitude is positively related to suicide, though this association seems to vary spatially. Secondly, Wang and colleagues [17] found that air pollution in China adversely affected depression symptoms, while neighborhood social capital seems to be a protective factor. Other than the social environment on a neighborhood level, close family also plays a crucial role in the development of mental disorders. For example, Guček and Selič [18] showed that exposure to intimate partner violence was a significant risk factor for the prevalence of depression, as were such life events as divorce. Xiao et al. [19] showed by means of structural equation models that housing conditions in Shanghai, China indirectly influenced migrants' mental health, whereas locals were directly affected. From a spatial planning point of view, the provision of environments supporting people's physical activity is central, as walkable areas reduce the risk of experiencing mental disorders. The study by Mayne et al. [20] addressed whether psychological distress is correlated with the walkability of the built environment at the zip code level in Sydney, Australia. Based on the absence of an association, the authors advised that health policies should focus on the personal level.

In conclusion, some of the papers in this special issue support the notion that environments can affect, in one way or the other, people's mental health. Although these studies advance our understanding of environment–health relations, there are several gaps in the context of the aforementioned contributions and in the literature on environmental health as a whole. For example, a key challenge for future research is how environmental exposures are assessed. It is traditionally assumed that residential location is the sole exposure source. However, the fragmentation of people's daily lives across numerous activity locations, as well as their residential mobility over the course of their lives, makes this approach questionable and calls for more comprehensive and dynamic exposure assessments [2]. Future research is advised to make the traversed environment central, as it might contribute to the onset of a mental disorder, and to integrate not only exposures at the actual place of residence but also those around past residential locations, as they may contribute to mental health disorders later in life.

Funding: This special issue is part of the NEEDS project (Dynamic Urban Environmental Exposures on Depression and Suicide, http://needs.sites.uu.nl/), which has received funding from the European Research Council (ERC) under the European Union's Horizon 2020 research and innovation program (grant agreement No. 714993).

Acknowledgments: My sincere thanks go to the authors who responded to the call for papers, as well as to the reviewers, whose support and critical and constructive comments on the manuscripts contributed enormously to the quality of this publication. Finally, I thank the IJERPH staff for editorial assistance throughout the preparation of this special issue.

Conflicts of Interest: The authors declare no conflict of interest. The founding sponsors had no role in the design of the study; in the collection, analyses, or interpretation of data; in the writing of the manuscript; or in the decision to publish the results.

References

1. Dzhambov, A.M.; Markevych, I.; Tilov, B.; Arabadzhiev, Z.; Stoyanov, D.; Gatseva, P.; Dimitrova, D.D. Pathways linking residential noise and air pollution to mental ill-health in young adults. *Environ. Res.* **2018**, *166*, 458–465. [CrossRef] [PubMed]
2. Helbich, M. Toward dynamic urban environmental exposure assessments in mental health research. *Environ. Res.* **2018**, *161*, 129–135. [CrossRef] [PubMed]
3. Bowler, D.E.; Buyung-Ali, L.M.; Knight, T.M.; Pullin, A.S. A systematic review of evidence for the added benefits to health of exposure to natural environments. *BMC Public Health* **2010**, *10*, 456. [CrossRef] [PubMed]
4. Allen, J.; Balfour, R.; Bell, R.; Marmot, M. Social determinants of mental health. *Int. Rev. Psychiatry* **2014**, *26*, 392–407. [CrossRef] [PubMed]
5. Evans, G.W. The built environment and mental health. *J. Urban Health* **2003**, *80*, 536–555. [CrossRef] [PubMed]
6. Vigo, D.; Thornicroft, G.; Atun, R. Estimating the true global burden of mental illness. *Lancet Psychiatry* **2016**, *3*, 171–178. [CrossRef]
7. Steel, Z.; Marnane, C.; Iranpour, C.; Chey, T.; Jackson, J.W.; Patel, V.; Silove, D. The global prevalence of common mental disorders: A systematic review and meta-analysis 1980–2013. *Int. J. Epidemiol.* **2014**, *43*, 476–493. [CrossRef] [PubMed]
8. Nichani, V.; Dirks, K.; Burns, B.; Bird, A.; Grant, C. Green space and depression during pregnancy: Results from the Growing Up in New Zealand Study. *Int. J. Environ. Res. Public Health* **2017**, *14*, 1083. [CrossRef] [PubMed]
9. Boers, S.; Hagoort, K.; Scheepers, F.; Helbich, M. Does residential green and blue space promote recovery in psychotic disorders? A cross-sectional study in the province of Utrecht, the Netherlands. *Int. J. Environ. Res. Public Health* **2018**, *15*, 2195. [CrossRef] [PubMed]
10. Schwartz, R.M.; Gillezeau, C.N.; Liu, B.; Lieberman-Cribbin, W.; Taioli, E. Longitudinal impact of Hurricane Sandy exposure on mental health symptoms. *Int. J. Environ. Res. Public Health* **2017**, *14*, 957. [CrossRef] [PubMed]
11. Schwartz, R.M.; Tuminello, S.; Kerath, S.M.; Rios, J.; Lieberman-Cribbin, W.; Taioli, E. Preliminary assessment of Hurricane Harvey exposures and mental health impact. *Int. J. Environ. Res. Public Health* **2018**, *15*, 974. [CrossRef] [PubMed]
12. Deisenhammer, E.A. Weather and suicide: The present state of knowledge on the association of meteorological factors with suicidal behaviour. *Acta Psychiatr. Scand.* **2003**, *108*, 402–409. [CrossRef] [PubMed]
13. Fernández-Niño, J.A.; Flórez-Garcia, V.A.; Astudillo-Garcia, C.I.; Rodriguez-Villamizar, L.A. Weather and Suicide: A Decade Analysis in the Five Largest Capital Cities of Colombia. *Int. J. Environ. Res. Public Health* **2018**, *15*, 1313. [CrossRef] [PubMed]

14. Kim, Y.; Ng, C.F.S.; Chung, Y.; Kim, H.; Honda, Y.; Guo, Y.L.; Lim, Y.-H.; Chen, B.-Y.; Page, L.A.; Hashizume, M. Air pollution and suicide in 10 cities in Northeast Asia: A time-stratified case-crossover analysis. *Environ. Health Perspect.* **2018**, *126*, 37002. [CrossRef] [PubMed]

15. Helbich, M.; de Beurs, D.; Kwan, M.-P.; O'Connor, R.C.; Groenewegen, P.P. Natural environments and suicide mortality in the Netherlands: A cross-sectional, ecological study. *Lancet Planet. Health* **2018**, *2*, e134–e139. [CrossRef]

16. Ha, H.; Tu, W. An ecological study on the spatially varying relationship between county-level suicide rates and altitude in the United States. *Int. J. Environ. Res. Public Health* **2018**, *15*, 671. [CrossRef] [PubMed]

17. Wang, R.; Xue, D.; Liu, Y.; Liu, P.; Chen, H. The relationship between air pollution and depression in China: Is neighbourhood social capital protective? *Int. J. Environ. Res. Public Health* **2018**, *15*, 1160. [CrossRef] [PubMed]

18. Guček, N.K.; Selič, P. Depression in intimate partner violence victims in Slovenia: A crippling pattern of factors identified in family practice attendees. *Int. J. Environ. Res. Public Health* **2018**, *15*, 210. [CrossRef] [PubMed]

19. Xiao, Y.; Miao, S.; Sarkar, C.; Geng, H.; Lu, Y. Exploring the Impacts of Housing Condition on Migrants' Mental Health in Nanxiang, Shanghai: A Structural Equation Modelling Approach. *Int. J. Environ. Res. Public Health* **2018**, *15*, 225. [CrossRef] [PubMed]

20. Mayne, D.J.; Morgan, G.G.; Jalaludin, B.B.; Bauman, A.E. Does walkability contribute to geographic variation in psychosocial distress? A spatial analysis of 91,142 members of the 45 and Up Study in Sydney, Australia. *Int. J. Environ. Res. Public Health* **2018**, *15*, 275. [CrossRef] [PubMed]

International Journal of
Environmental Research and Public Health

MDPI

Article

Green Space and Depression during Pregnancy: Results from the *Growing Up in New Zealand* Study

Vikram Nichani [1,*]**, Kim Dirks** [1] **, Bruce Burns** [2] **, Amy Bird** [3] **and Cameron Grant** [3,4,5]

[1]   Section of Epidemiology and Statistics, School of Population Health, University of Auckland, Auckland 1142, New Zealand; k.dirks@auckland.ac.nz
[2]   School of Biological Sciences, University of Auckland, Auckland 1142, New Zealand; b.burns@auckland.ac.nz
[3]   Centre for Longitudinal Research he Ara ki Mua, School of Population Health, University of Auckland, Auckland 1142, New Zealand; a.bird@auckland.ac.nz (A.B.); cc.grant@auckland.ac.nz (C.G.)
[4]   Department of Pediatrics: Child and Youth Health, School of Medicine, University of Auckland, Auckland 1142, New Zealand
[5]   General Pediatrics, Starship Children's Hospital, Auckland District Health Board, Auckland 1023, New Zealand
*   Correspondence: vnic041@aucklanduni.ac.nz; Tel.: +64-(0)-9-923-6722

Received: 25 July 2017; Accepted: 12 September 2017; Published: 18 September 2017

Abstract: *Background*: Antenatal depression is an important contributor to poor maternal health experienced by some women. This study aimed to determine whether exposure to green space during pregnancy is associated with less depression, and whether this association is moderated by relevant factors, such as age, education, self-identified ethnicity, physical activity, residential rurality, and socioeconomic status. *Methods*: Health data were sourced from the cohort study "*Growing Up in New Zealand*" comprised of 6772 participants. Green space was estimated based on the proportion of green space within the Census Area Unit. Adjusted logistic mixed effect models were used to investigate the association between green space and antenatal depression after controlling for confounding variables. *Results*: Maternal exposure to green space were not associated with lower odds of antenatal depression. Indications of effect modifications due to relevant factors were not observed. *Conclusions*: This study did not determine an association between access to green space (measured based on the distance to the nearest green space) and antenatal depression. Therefore, a link between green space and antenatal depression was not established. For that reason, ensuring residential areas contain adequate green space may or may not be helpful in preventing antenatal depression and adverse health outcomes associated with this depression. More studies focusing on pregnant women in a range of social contexts, and considering both exposure and access to green space, are warranted to determine the relationships between green space and antenatal depression.

Keywords: antenatal depression; green spaces; census area units; geographic information systems; multilevel data

---

1. Introduction

According to the World Health Organization, the most common mental health disorder affecting adults in the general population is depression [1]. In the United States of America (USA), billions of dollars each year are spent on depression, attributable to direct medical costs (i.e., medical services and prescription drug costs), suicide-related mortality, and workplace costs (i.e., costs associated with an absence from work and reduced productivity) [2]. Depression is clinically diagnosed by the presence of unhappiness; feelings of guilt; tiredness; and lack of appetite, sleep, concentration, and pleasure [1]. The prevalence rate of depression amongst adults in the general population is country-specific, ranging from 11% for low and medium income countries to 15% for high income countries [3]. Depression can manifest as a chronic condition, and, if left untreated, leads to loss of productive life years during the life course of the affected individual [1,4]. In 2001, depression was regarded as the fourth leading cause of disability [5]. By 2020, depression is projected to become the second most important factor associated with disability [5]. In the most severe cases, depression culminates into suicide [1].

Globally, variable prevalence rates of antenatal depression (defined as "depression during pregnancy") have been observed, ranging from 12% to 20% of pregnant women [6]. In New Zealand, it is estimated that 15% of pregnant women (hereafter referred to as women) suffer from mental disorders, including anxiety and depression [7]. Antenatal depression is an important contributor to the adverse pregnancy outcomes of low birth weight, preterm birth, small for gestational age, smaller head infants, and adverse child health outcomes of low Apgar scores and infant mortality [8,9]. Additionally, antenatal depression is one of the important etiological factors responsible for the development of postnatal depression [10,11]. Postnatal depression in mothers is linked to higher cognitive, behavior, and interpersonal problems in their children [12].

One environmental intervention that can combat antenatal depression in women is exposure to green space [13]. The United States Environmental Protection Agency has defined green space as "land that is partly or completely covered with grass, trees, shrubs, or other vegetation" [14]. It has been observed that women living in space that is more green are less likely to develop antenatal depression [13]. At the same time, exposure to green space increases the levels of physical activity in women [13] and children [15]. In this context, recent studies have indicated that the development of large green space areas are frequently preferred in landscape planning stages [16–18]. The beneficial effects of green space exposure during pregnancy are more evident in women of lower socioeconomic status [13], those with low levels of education [19–22], or in those living in highly deprived areas [21]. More specifically, recent studies on green space and pregnancy outcomes indicate that women with low or medium levels of education deliver higher birth weight infants compared with women with high levels of education [19–22]. The association of green space exposure with higher birth weight is stronger among women residing in the most deprived areas compared with women residing in the moderate or least deprived areas [21]. Thus, a moderating effect of socioeconomic status in the association between green space and pregnancy outcomes has been demonstrated. At the same time, data from some general population studies are suggestive of a moderating effect of age in the association between green space and mental health [23,24], and a moderating effect of residential rurality in the association between green space and general health [25]. More exactly, general population studies on the associations between exposure to green space and mental health outcomes have demonstrated that green space is beneficial for the mental health of people within specific age groups (e.g., 18–24, <30, and 31–50 years) [23,24]. A study of the association between exposure to green space and health outcomes amongst general population adults has shown that people living in greener areas have better health outcomes (e.g., morbidity symptoms and perceived general health status) in comparison with those living in low-green areas; these associations are seen mainly in people living in the slightly urban/moderately urban/nonurban areas [25]. In the general population, the association between green space with mental health outcome of psychological distress is more prominent in people who are physically active [26]. That is, people who live in areas of high green space and who are physically active are less likely to develop psychological distress in comparison with people who live in low green areas and who are physically inactive [26]. It is also known that green space is associated with better mental health in women, at least in part through increased participation in physical activity [13]. One limitation that appears in previous general population studies [27,28], and those focused specifically on women [13] and green space in relation to mental health, is that the studies have not accounted for nor controlled for self-selection bias in the regression analyses.

We aimed to investigate whether exposure to green space for women was associated with a lower likelihood of antenatal depression after accounting for confounders, including socioeconomic status and the length of stay at their current residence, used as a surrogate for self-selection bias.

We also aimed to investigate whether the effect of green space exposure on antenatal depression varied between different age and ethnic groups, low/medium/high levels of education, urban/rural groups, low/medium/high area deprivation groups, and for physically active groups. We sourced health data from a cohort that sampled women of diverse ethnicity and socioeconomic status, and gathered data on physical activity, so that the effect of modifications of demographic and residential factors and physical activity on the relationship between green space exposure and antenatal depression could be investigated.

2. Materials and Methods

2.1. Study Source

Data for this study were sourced from mothers who were participants in the *Growing Up in New Zealand* study, a longitudinal pre-birth cohort study of 6853 children and their parents who are residents of the Auckland, Manukau, or Waikato regions of New Zealand [29]. The study region, covered by three adjacent District Health Boards of Auckland, Counties Manukau and Waikato, represented 11% of the live births in New Zealand, from March 2009 to May 2010 [29]. *Growing Up in New Zealand* recruited 6822 women for the first data collection wave, called the "antenatal wave", whereby data were collected through face-to-face interviews with women [29]. Participants of the *Growing Up in New Zealand* study were interviewed prior to the birth of their child or children, as well as after the birth [29]. Written informed consent was obtained from mothers for their participation, as well as of their unborn children [29]. As part of the antenatal wave, data that described demographics, health behaviors and history, and household characteristics were collected from mothers [29].

2.2. Estimation of Exposure to Green Spaces

The assessment of green space was performed based on the proportion of green space within a given census area unit (CAU). CAUs in New Zealand are the second smallest geographical units consisting of populations of 3000 to 5000 [30]. Statistics New Zealand define census area units as "non-administrative areas that are in between mesh blocks and territorial authorities in size" [30]. Our method of assessment of green space was similar to that used in previous studies in New Zealand [31–33], namely, by dividing green spaces within CAUs into different quartiles based on the percentage of green space in the CAUs. Our definition of green space included green areas, such as parks, beaches, urban parklands/open spaces, forests, grasslands, and croplands, but excluded private gardens. Other non-green areas (e.g., built-up areas (e.g., commercial, industrial, and residential buildings), space used to support transport infrastructure (e.g., roads, rail-yards, and airport runways) and water bodies (e.g., rivers and lakes)) were also excluded from our measure of green space. Data on green space for the study region were sourced from the Auckland Council [34] and the Waikato District Council. We supplemented data on green space from two Councils with data on green space from the New Zealand Land Cover Database (LCDB) of the Land Resource Information Systems portal [35]. The procedure of combining data on green space provided more attributes than using data from a single source. Green space data from the Waikato District Council had a scale of 1:50,000 and an accuracy of 90.0%. The LCDB data had a scale of 1:50,000 and an accuracy of 93.9% [35]. As the relationship between green space and antenatal depression was non-linear, the green space variable was utilized as a categorical variable. We took the 25th, 50th, and 75th percentiles as the break points for the categorization of green space. The utilization of those percentiles resulted in green space being categorized in our study as low (0% to <12%), medium (12% to <21%), high (21% to <38%), and very high (38% to 100%). The Aeronautical Reconnaissance Coverage Geographic Information System (ArcGIS) Version 10.3 (Environmental Systems Research Institute, Redlands, CA, USA) was used to perform the green space analyses.

2.3. Covariates

The covariates used were age (categorized as <20, 20–24, 25–29, 30–34, 35–39, and ≥40 years), ethnicity (defined as self-identified ethnicity and categorized into European, Māori (New Zealand's indigenous population), Pacific, Asian, Middle Eastern/Latin-American/African, and New Zealander/ Other), educational attainment (defined as the highest level of education attained and categorized as no secondary school qualification, secondary school, diploma certificate, bachelor's degree, and higher degree), employment status (defined as status in labor force service and categorized as employed, unemployed, student, and not in work force), area deprivation (defined as "the New Zealand Deprivation Index 2006 [NZDep2006]") which is obtained by combining a set of variables collected during the 2006 national census (e.g., income, home ownership, living space, access to telephone, and access to car) [36], and categorized into deprivation deciles of low (deciles [1–3]), medium (deciles [4–7]), and high (deciles [8–10]), smoking status (defined as smoking of cigarettes during pregnancy, and categorized as yes or no), alcohol consumption (defined as consumption of alcohol during pregnancy, and categorized as no drinking during pregnancy or any drinking during pregnancy), relationship status with biological father (defined as social relationship status with biological father, and categorized as no relationship; dating, not cohabiting; cohabiting; and married or civil union), parity (defined as the number of pregnancies and categorized as first or subsequent), residential rurality (defined as residence in urban or rural areas), physical activity during and after the first trimester of pregnancy (defined as participation in recommended levels of physical activity of at least 150 min per week [37], and categorized as yes or no), pre-pregnancy general health status (defined as general health status during pre-pregnancy period and categorized as poor/fair, good, very good, or excellent), and the length of stay at the current residence (measured in years and described below) [29].

2.4. Self-Selection Bias

The importance of self-selection bias has been recognized in studies on green space and health. For example, a systematic review of the association between measures of the built environment (e.g., parks and public open spaces) and physical activity amongst adults from the general population has identified that neighborhood self-selection is likely to be a confounder of the association between measures of the built environment and physical activity [38]. One way to reduce the creation of biased estimates while determining the association between measures of the built environment and physical activity is to statistically control for neighborhood self-selection [39].

It is likely that the association between exposure to green spaces and depression amongst general population adults is confounded by the process of neighborhood self-selection [40]. This choice could, at least in part, explain an association between exposure to green space and depression amongst general population adults [40]. Consequently, the length of stay at the current residence had to be considered while examining the association between green space and antenatal depression in this study.

Consistent with previous research studies on exposure to green space and mental health outcomes amongst general population adults [40,41], we used the variable "length of stay at current residence" as a surrogate measure for neighborhood self-selection. This is taken into account because a minimum length of time (at least one year of stay at the current residence [42]) is needed before the beneficial effects of exposure to green space on mental health become evident. In the *Growing Up in New Zealand* study, the variable "length of stay at current residence" was described as the number of years that the women had lived in their current residence [43], framed as "How long have you lived in this current home?" and specifying the number of months, or number of years, or both, that they lived in their current home [43].

2.5. Dependent Variable

The dependent variable utilized in this study was the Edinburgh Postnatal Depression Scale (EPDS) as it is the most common screening instrument used for the detection of antenatal and post-natal depression [44,45]. The EPDS questionnaire consists of 10 questions that extract in-depth information on antenatal or postnatal depression [44,45]. Each question has four responses (e.g., Yes, most of the time; Yes, quite often; Not very often; and No, not at all) and a rating score of 0–3 points, with the maximum calculated total score for any individual being 30 points [46]. Both the validity and reliability of the EPDS have been demonstrated for its usage in diverse cultures [47–49]. For women who had limited ability to speak English, or those who could not speak the English language, an interpreter was available so that the EPDS questionnaires could be administered.

In the *Growing Up in New Zealand* study, women were asked to recollect information over the past seven days while answering the following ten questions of the EPDS questionnaire: (1) I have been able to laugh and see the funny side of things, (2) I have looked forward with enjoyment to things, (3) I have blamed myself unnecessarily when things went wrong, (4) I have been anxious or worried for no good reason, (5) I have felt scared or panicky for no very good reason, (6) Things have been getting on top of me, (7) I have been so unhappy that I have had difficulty sleeping, (8) I have felt sad or miserable, (9) I have been so unhappy that I have been crying, and (10) The thought of harming myself has occurred to me. After gathering information from the EPDS questionnaires, the categorizations of antenatal depression and non-depression in the *Growing Up in New Zealand* study were made at a cut-off value of a total score of 13 points. At the threshold value of 13 points, the sensitivity for antenatal depression is 0.83 and specificity for antenatal depression is 0.90 [46]. A total score of ≥13 points was considered to be associated with a high risk for antenatal depression [43]. For the analyses relating to the current study, the EPDS scores for women were dichotomized as 0 (absence of antenatal depression) or 1 (presence of antenatal depression) using the same cut-off value of 13 points.

2.6. Statistical Analyses

The final sample of 6772 mothers in this study from the original cohort of 6822 mothers was created by restricting analyses to those women for whom each geocoded CAU number was available for analysis. There were 613 CAUs within the study regions. On average, 11 respondents resided in each CAU. If the clustering of respondents in CAUs of study regions is not accounted for within the regression models, the standard errors of the contextual effect estimates are likely to be downwardly biased. Multilevel designs are ideally suited to the analysis of neighborhood effects, such as green space, within different sized CAUs, by simultaneously analyzing individual and neighborhood-level variables, whilst accounting for the non-independence in the data [50]. As the proportion of missing values in the variables was low, we decided to perform a complete case analysis. We conducted a multilevel mixed model for a green space-depression association using district health board of the maternal domicile as a random effect. As the EPDS score could not be normalized, we used EPDS as a binary variable. The length of stay at the current residence was modeled as a continuous variable and the green space-depression association judged based on logistic regression by observing odds ratios (ORs) and their 95% confidence intervals (CIs). Firstly, we conducted univariate regression analyses (Null Model) to determine whether exposure to green space by itself was a significant predictor of antenatal depression. Subsequently, confounders were added to the null models one by one to determine whether exposure to green space remained a significant predictor of antenatal depression. Interaction tests determined the possibility of interactions between green space and relevant variables. The final regression model had all independent variables, including green space, and interaction terms if they led to a better fit in the likelihood ratio tests. A total of seven mixed models were developed in Stata Version 14 (Stata Corporation, College Station, TX, USA) for investigating the association between green space and antenatal depression.

3. Results

3.1. Exposure to Green Spaces

The median area within CAUs of the Auckland and Counties Manukau District Health Board regions was 1.62 km^2 (*n* = 413 CAUs). On the other hand, the median area within CAUs of the Waikato District Health Board region was 6.64 km^2 (*n* = 200 CAUs). The mean (standard deviation (SD)) percentage of green space in the CAUs for the Auckland and Counties Manukau District Health Board regions was 38% (32%). Correspondingly, the mean (SD) percentage of green space in the CAUs of the Waikato District Health Board region was 65% (34%).

Educational attainment (χ^2 = 47.01; *p* < 0.0001) and self-identified ethnicity (χ^2 = 352.00; *p* < 0.0001) were significantly different among population subgroups in terms of their exposure to green space. Women who had acquired diploma level education qualifications were exposed to higher surrounding greenness within CAUs of their residence than those with other levels of educational qualification (diploma = 32.65% (95% CI = 31.42–33.88%), no secondary school = 31.05% (95% CI = 28.70–33.40%), secondary school = 30.83% (95% CI = 29.50–32.16%), bachelor's degrees = 30.63% (95% CI = 29.19–32.07%), and higher degrees = 28.00% (95% CI = 26.38–29.63%)). European women were exposed to higher surrounding greenness within CAUs of their residence than non-European women (European = 36.14% (95% CI = 35.10–37.17%), Māori = 31.08% (95% CI = 29.43–32.73%), Pacific = 21.90% (95% CI = 20.93–22.86%), Asian = 21.31% (95% CI = 20.14–22.50%), Middle Eastern/Latin-American/African = 26.58% (95% CI = 23.12–30.05%), and Other or New Zealander = 34.95% (95% CI = 28.42–41.48%)).

Exposure to green space was also significantly different for population subgroups based on the level of area deprivation (F = 172.79; *p* < 0.001). Mothers residing in low deprivation areas were exposed to higher levels of green space in comparison to those residing in medium and high deprivation areas. More specifically, the proportion of green space available to women residing in low, medium, and high deprivation areas was 39.93% (95% CI = 38.33–41.53), 31.85% (95% CI = 30.68–33.02), and 24.23% (95% CI = 23.50–24.96), respectively.

3.2. Descriptive Statistics and Bivariate Analyses

The mean (SD) of the maternal age of mothers enrolled in the *Growing Up in New Zealand* study was 30 (6) years, with most of these women, 93%, residing in urban areas at the time of cohort enrollment (Table 1). The mean (SD) duration of the length of stay at the current residence for mothers was 4 (6) years (Table 1). Fifteen per cent of these mothers experienced depression during the antenatal period, and the proportion of the study cohort experiencing antenatal depression who resided in areas of low, medium, high, and very high green space were 14.68%, 17.32%, 17.45%, and 15.22%, respectively.

Significant differences were identified among population subgroups defined by NZDep2006 in terms of length of stay at the current residence (F = 1.62; *p* < 0.0001). The mean lengths of stay at the current residence for women residing in low, medium, and high deprivation areas were 3.89, 3.96, and 4.86 years, respectively. In contrast, the length of stay at the current residence was not associated with green space (F = 1.15; *p* = 0.10), residential rurality (F = 0.95; *p* = 0.68), physical activity during (F = 1.15; *p* = 0.09) and after the first trimester of pregnancy (F = 0.93; *p* = 0.74), and antenatal depression (F = 1.00; *p* = 0.50).

Table 1. Demographics and other characteristics of antenatal mothers and bivariate analyses for antenatal depression used to assess the effect of green space on antenatal depression from the *Growing Up in New Zealand* study cohort.

Variables (*n* = 6772) *	Descriptives *n* (%) or Mean (SD)	Bivariate Analyses Test Statistic (*p*-Value for Chi-Square or F Test)
Demographics		
Age (years), *n* (%)		
<20	325 (5)	147.31 (<0.001)
20–24	992 (15)	
25–29	1651 (24)	
30–34	2108 (31)	
35–39	1411 (21)	
≥40	285 (4)	
Education, *n* (%)		
No secondary school	485 (7)	123.47 (<0.001)
Secondary school	1610 (24)	
Diploma	2068 (30)	
Bachelor's degree	1532 (23)	
Higher degree	1058 (16)	
Employment status, *n* (%)		
Employed	3636 (54)	93.83 (<0.001)
Unemployed	543 (8)	
Student	455 (7)	
Not in workforce	1822 (27)	
Self-identified ethnicity, *n* (%)		
European	3576 (53)	202.76 (<0.001)
Māori	933 (14)	
Pacific	1001 (15)	
Asian	1002 (15)	
Middle Eastern/Latin-American/African	145 (2)	
Other or New Zealander	96 (1)	
Relationship status, *n* (%)		
No relationship	125 (2)	115.04 (<0.001)
Dating, not cohabiting	278 (4)	
Cohabiting	2312 (34)	
Married or civil union	4038 (60)	
Parity, *n* (%)		
First born	2833 (42)	0.68 (0.41)
Subsequent	3932 (58)	
Health behaviors and health		
Smoking, *n* (%)		
No	5473 (81)	101.53 (<0.001)
Yes	656 (10)	
Alcohol consumption, *n* (%)		
No drinking during pregnancy	4851 (72)	9.12 (0.003)
Any drinking during pregnancy	1904 (28)	
Physical activity during the first trimester of pregnancy, *n* (%)		
No	4243 (63)	2.91 (0.088)
Yes	1890 (28)	
Physical activity after the first trimester of pregnancy, *n* (%)		
No	4700 (69)	0.64 (0.42)
Yes	1433 (21)	

Table 1. *Cont.*

Variables (*n* = 6772) *	Descriptives *n* (%) or Mean (SD)	Bivariate Analyses Test Statistic (*p*-Value for Chi-Square or F Test)
Demographics		
General health status, *n* (%)		
Poor or fair	698 (10)	168.79 (<0.001)
Good	2306 (34)	
Very good	2382 (35)	
Excellent	1372 (20)	
Antenatal depression, *n* (%)		
No	5141 (76)	-
Yes	992 (15)	
Household characteristics		
Area deprivation (NZDep2006), *n* (%)		
≤3: low	1684 (25)	90.83 (<0.001)
4–7: medium	2471 (36)	
8–10: high	2615 (39)	
Length of stay at current residence, mean (SD)		
Length of stay (years)	4 (6)	1.00 (0.50)
Residential rurality, *n* (%)		
Urban	6325 (93)	3.52 (0.06)
Rural	447 (7)	
District health board region, *n* (%)		
Auckland	2421 (36)	53.70 (<0.001)
Manukau	2526 (37)	
Waikato	1825 (27)	
Green space percentage in census area units		
Green space percentage, *n* (%)		
Low (0–<12%)	1672 (25)	6.87 (0.08)
Medium (12–<21%)	1652 (24)	
High (21–<38%)	1764 (26)	
Very High (38–100%)	1684 (25)	

* The proportion of missing data for each variable is from 0.0 to 9.5%.

3.3. Interaction Tests for the Association of Green Space with Antenatal Depression

We did not find any interactions between green space and relevant factors while determining the association between green space and antenatal depression (area deprivation ($p = 0.07$), physical activity during the first trimester of pregnancy ($p = 0.07$), age ($p = 0.09$), physical activity after the first trimester of pregnancy ($p = 0.45$), residential rurality ($p = 0.75$), self-identified ethnicity ($p = 0.83$), and educational attainment ($p = 0.94$). We did not perform any subgroup analyses for groups of physical activity after the first trimester of pregnancy, residential rurality, self-identified ethnicity, and educational attainment. The *p*-values for interactions between green space and relevant factors, such as area deprivation, physical activity during the first trimester of pregnancy, and age, were close to significance; therefore, we performed subgroup analyses for different areas of deprivation, physical activity during the first trimester of pregnancy, and age groups. The results of these subgroup analyses were similar to those of the main analyses, and indicative of lacking associations between green space and antenatal depression (data not shown).

3.4. Main Analyses for the Association of Green Space with Antenatal Depression for the Entire Cohort

Univariate logistic regression analysis revealed that exposure to medium, high, or very high levels of green space was not associated with antenatal depression (unadjusted OR medium green space = 1.11

(95% CI = 0.91–1.36); unadjusted OR high green space = 1.12 (95% CI = 0.92–1.36); and unadjusted OR very high green space = 0.95 (95% CI = 0.77–1.17)) (Null Model). Similarly, no association was found between exposure to green space and antenatal depression after accounting for all confounders. That is, the fully adjusted multivariate regression analyses (Model 7) showed that exposure to medium, high, or very high levels of green space was not associated with antenatal depression (adjusted OR medium green space = 1.10 (95% CI = 0.89–1.35); adjusted OR high green space = 1.15 (95% CI = 0.94–1.41); and adjusted OR very high green space = 1.21 (95% CI = 0.96–1.52)) (Table 2).

Table 2. Multilevel analyses for maternal exposure to green space during pregnancy and odds of antenatal depression for the entire cohort used to assess the effect of green space on antenatal depression from the *Growing Up in New Zealand* study cohort.

Green Space Percentage	Association between Maternal Exposure to Green Space and Antenatal Depression (n = 6772)			
	Model 1 [a]	Model 2 [b]	Model 3 [c]	Model 4 [d]
	OR (95% CI)	OR (95% CI)	OR (95% CI)	OR (95% CI)
Low	1.00	1.00	1.00	1.00
Medium	1.11 (0.91–1.36)	1.13 (0.92–1.39)	1.12 (0.91–1.37)	1.10 (0.90–1.35)
High	1.12 (0.92–1.36)	1.13 (0.93–1.39)	1.13 (0.92–1.39)	1.12 (0.92–1.38)
Very High	0.95 (0.77–1.17)	1.20 (0.97–1.49)	1.20 (0.96–1.48)	1.22 (0.98–1.51)
Green space percentage	Model 5 [e]	Model 6 [f]	Model 7 [g]	
	OR (95% CI)	OR (95% CI)	OR (95% CI)	
Low	1.00	1.00	1.00	
Medium	1.11 (0.90–1.36)	1.10 (0.90–1.36)	1.10 (0.89–1.35)	
High	1.14 (0.93–1.40)	1.14 (0.93–1.40)	1.15 (0.94–1.41)	
Very High	1.22 (0.98–1.51)	1.20 (0.97–1.49)	1.21 (0.96–1.52)	

[a] = unadjusted univariate; [b] = adjusted for maternal age and self-identified ethnicity; [c] = adjusted for maternal age, self-identified ethnicity, smoking, and alcohol consumption during pregnancy; [d] = adjusted for maternal age, self-identified ethnicity, smoking, alcohol consumption during pregnancy, pre-pregnancy general health status, physical activity during the first trimester and remainder of pregnancy; [e] = adjusted for maternal age, self-identified ethnicity, smoking, alcohol consumption during pregnancy, pre-pregnancy general health status, physical activity during the first trimester and remainder of pregnancy, relationship status, and parity; [f] = adjusted for maternal age, self-identified ethnicity, smoking, alcohol consumption during pregnancy, pre-pregnancy general health status, physical activity during the first trimester and remainder of pregnancy, relationship status, parity, education, and employment status; [g] = adjusted for maternal age, self-identified ethnicity, relationship status, parity, smoking, alcohol consumption, general health status, physical activity during the first trimester and remainder of pregnancy, education, employment status, residential rurality, area deprivation, and length of stay at current residence.

4. Discussion

4.1. Main Findings

The range of green space exposure for the participants in this study was similar to the range of green space exposure found in previous studies conducted in New Zealand [31–33]. Additionally, in this study, we observed a socioeconomic gradient in exposure to green space. That is, green space decreased as area deprivation increased. This result is not different from what has been observed in previous European studies that have investigated the relationship between exposure to green space for women and adverse pregnancy outcomes of low birth weight and preterm birth [19,21]. Indeed, similar results have been observed in a previous general population study from New Zealand; a study of the association between exposure to green spaces and cause-specific mortality in adults from the general population in New Zealand revealed a socioeconomic gradient in exposure to green space [32]. In the aforementioned New Zealand study, one unit increase in deprivation score resulted in a decrease in green space of 11% [32].

The findings of the current study suggest that exposure to green space for the entire cohort of women is not beneficial in decreasing the odds of antenatal depression. Additionally, this study could not demonstrate that associations of green space with antenatal depression were stronger for specific population subgroups.

4.2. Comparison of Study Results with Previous Studies

We add to the already existing literature on green space and antenatal depression by exploring the issue of whether green space exposure does or does not reduce the odds of depression during pregnancy. Building on previous research on green space-depression relationships in women from England [13], our New Zealand-based study investigated the green space-depression association for a cohort (*Growing Up in New Zealand*) whilst controlling for confounders, including socioeconomic status, and the length of stay at the current residence. This is the first study in New Zealand that has accounted for self-selection bias for the association between exposure to green space and antenatal depression utilizing the same cohort. The overall relationships in this study were not in the predicted directions. We expected green space to be associated with a lower likelihood of antenatal depression for the entire cohort of women. More importantly, we expected women who were physically active to be less depressed than those who were not physically active. Similarly, we expected women with low levels of education to be less depressed than those with medium or high levels of education.

McEachan and colleagues utilized the Born in Bradford cohort sample to examine the relationships between exposure and access to green space for women and the odds of antenatal depression [13]. In the Born in Bradford Study, depression amongst the study participants was measured through the administration of the General Health Questionnaires (GHQs). More specifically, a total of four questions were answered on a four-point Likert scale (i.e., 0 to 3) by the study participants. A binary variable, representing the likelihood of depression, was constructed and categorized as "depression" (scores of 0 on all four questions) and "non-depression" (score of 1 on at least 1 question) [13]. In contrast with our study, the Born in Bradford study showed that women residing in the greenest areas had a lower likelihood of antenatal depression when compared to those residing in the least green areas (OR = 0.82 (95% CI = 0.69–0.98)) [13]. Additionally, the association between exposure to green space and antenatal depression in the Born in Bradford cohort was independently identified for women who had attained only low levels of education, as well as women who were physically active throughout their pregnancy. Women with only low levels of education and living in the greenest areas were *less* likely to develop antenatal depression than those living in least green areas (OR = 0.74 (95% CI = 0.59–0.94)) [13]. Those who had achieved the recommended levels of physical activity during pregnancy (at least 150 min of moderate physical activity on average a week [51]) and lived in the greenest areas were *less* likely to develop antenatal depression than those living in the least green areas (OR = 0.63 (95% CI = 0.41–0.97)) [13]. The Born in Bradford study also showed that the relationship between access to green space and antenatal depression was significant for the study participants. Women who lived within 300 m of a major green space were 13% less likely to develop antenatal depression than those who lived >300 m of a major green space [13]. That is, the fully adjusted multivariate regression analyses revealed that women who lived within 300 m of a major green space had lower odds for development of antenatal depression in comparison to those who lived >300 m of a major green space (OR = 0.87 (95% CI = 0.77–0.99)) [13]. No significant green space-depression associations were identified for specific pregnant population subgroups (e.g., those with low levels of education or those who were physically active) while considering access to green space [13]. Some explanations can be given for the contrasting results between our study and Born in Bradford study. One possible explanation is the lack of environmental variation in green space in New Zealand. In New Zealand, cities generally provide a high amount of green space [32,37,52,53]. The high levels of exposure to green spaces due to outdoor vacationing (camping) may be responsible for a lack of variation in exposure to green spaces in New Zealand [27]. Richardson and colleagues went so far as to suggest that, unlike in other places such as in England, green space is not an important determinant of health in New Zealand as green space is everywhere, leading to a lack of variation in exposure [32]. Another explanation for the differences in results between our study and the Born in Bradford study is the inclusion of the neighborhood self-selection variable in the regression analyses. We included the length of stay at the current residence variable as a surrogate for neighborhood self-selection in our regression models whereas the Born in Bradford study did not [13]. However, over the last decade, there has been a housing shortage in Auckland, in particular, due to a failure in the housing market. Therefore, the length of stay at the current residence may not be an ideal surrogate measure of neighborhood self-selection.

We can also compare our study results to the results of prior general population studies on green space-mental health carried out both within and outside of New Zealand. Our lack of association between exposure to green space and antenatal depression for the entire cohort of women is consistent with the results of a few general population studies on green space and mental health [24,28,54], but not with others [26,27,31,40]. One cross-sectional study assessed the association between exposure to green space and poor mental health in adults from the general population in New Zealand [31]. Poor mental health was judged on the basis of the presence of a short form-36 mental health score in the lowest quartile [31]. This study showed that exposure to green space was associated with lower odds of poor mental health. That is, people residing in the greenest CAUs were *less likely* to report poor mental health than those residing in the least green CAUs (OR = 0.81 (95% CI = 0.66–1.00)) [31]. Another general population study in New Zealand examined the association between exposure/access to green space and mental health amongst adults [27]. In this general population study, mental health was assessed through a number of anxiety/mood disorder treatment counts, and this number included people who had received secondary treatments for mental health disorders, or who had received subsidized prescription drugs for anxiety/mood disorders, or those diagnosed for mental health disorders on the basis of positive laboratory tests for lithium [27]. This study showed that both increases in access to green space (defined as decreased distances to green spaces) and increases in exposure to green spaces in residential environments were associated with decreased anxiety/mood disorder treatment counts. Each decrease in distance to the nearest green space by 100 m was associated with a 3% reduction in anxiety/mood disorder treatment counts [27]. Additionally, an increase in the proportion of greenness in 3 km buffers around the residence by 1% resulted in a decrease in anxiety/mood disorder treatment counts by 4% [27]. An Australian study investigated the association between exposure to green space and mental health amongst adults from the general population [26]. In this green space-mental health study, the participant's mental health was assessed through the administration of the psychological distress questionnaire which indicated psychological distress [26]. Each questionnaire consisted of 10 questions which were answered on five-point Likert scales (e.g., 1 = none of the time to 5 = all of the time) [26]. Total scores of ≥22 were indicative of the presence of psychological distress [26]. Adults residing in the greenest neighborhoods were *less likely* to develop psychological distress in relation to those residing the least green neighborhoods (OR = 0.83 (95% CI = 0.76–0.92) [26]. Additionally, people living in the greenest neighborhoods were found to be *less* sedentary compared with people living in the least green neighborhoods (OR = 0.81 (95% CI = 0.77–0.87)) [26]. A recent general population cross-sectional study explored the association between exposure to green space and mental health of residents in the state of Wisconsin, USA [40]. In this study, the mental health of the study participants was accessed through the 42-item depression, anxiety, and stress scale, which indicated symptoms of depression, anxiety, and stress [40]. The results showed that green space was associated with less depression, anxiety, and stress. A 25% increase in the coverage of green space in the residential environment was associated with a decreases in the scores for depression, anxiety, and stress by 1.379, 0.427, and 0.735 points, respectively [40]. A British longitudinal design study conducted in general population adults investigated the association between exposure to green space and mental health disorder of minor psychiatric morbidity [55]. The general health scores obtained from 12-item GHQs were used to assess mental health statuses with higher scores indicating presence of minor psychiatric morbidity [55]. The general health construct within a 12-item GHQ examined a range of questions on mental health issues such as concentration, insomnia, lack of confidence, self-worthiness, happiness, and depression [55]. Analysis of the green space-mental health associations showed that green space exposure was beneficial for women aged >41 years of age resulting in a reduction in the mean GHQ score [55]. The association of green space with minor psychiatric morbidity persisted until old age for women, though beneficial only for exposure to moderate levels of green space, defined as 34–66% green space coverage in the small geographical areas of Wards [55]. In men, the associations of moderate (34–66% green space coverage) and high levels (67–100% green space coverage) of green space with minor psychiatric morbidity were seen in early adulthood appearing at 30 years, and these peak at 41–45 years. These associations remained till 60 years, after which they were not seen in old age [55].

A study of the associations between exposure to blue and green spaces (in terms of the visibility of blue and green spaces) and psychological distress in adults in New Zealand failed to establish any connection between exposure to green space and psychological distress [28]. In the aforementioned New Zealand study, investigators assessed the mental health of the study participants through psychological distress scale scores, with higher scores indicating presence of psychological distress [28]. This study showed that blue space visibility was associated with lower scores on psychological distress scales ($\beta = -0.28$; p-value < 0.001), but green space visibility was not associated with lower scores on psychological distress scales ($\beta = -0.09$; p-value = 0.455) [28]. This meant that each 10% increase in blue space and green space visibility was associated with a decrease in psychological distress score by 0.28 and 0.09 points, respectively [28]. However, the investigators of this study did not include private gardens in their classification of green spaces, or assess the quality of green spaces, nor control for the length of stay at current residence in the regression analysis [28]. It is possible that their inclusion could have resulted in significant associations between green space visibility and psychological distress [28]. A general population cohort study investigating the association between access to green space qualities and poor mental health has also been carried out through the administration of a 12-item GHQs in Swedish adults. This study showed that access to green space qualities for the entire cohort (OR men = 1.10 (95% CI = 0.7–1.60); OR women = 1.10 (95% CI = 0.80–1.60)) was not associated with the development of poor mental health [54]. However, women who had access to green space and were also physically active were *less* likely to develop poor mental health, in comparison to women without access to green space and who were physically inactive (OR women = 0.30 (95% CI = 0.10–0.90)) [54]. Study limitations included the use of a non-validated measure of physical activity and the use of self-reported measures of access to green space qualities [54].

Considering all general population studies, it is possible to say that the directional trends for the associations of green space with mental health outcomes across several studies is essentially mixed. That is, there is a lack of consensus among general population studies conducted within and outside New Zealand on the effect of green space on mental health outcomes. Most general population studies conducted within and outside New Zealand suggest that exposure to green space improves mental health outcomes [26,27,31,40], whilst some others have found no association between exposure to green space and mental health outcomes [28,54]. One possible reason for the lack of consensus between different mental health studies may be variation in the definition of mental health. For example, mental health has been defined in various ways in general population studies, such as poor mental health [31,54], minor psychiatric morbidity [55], psychological distress [26,28], anxiety/stress/depression [24,40], and counts of anxiety/mood disorder [27]. Self-assessment of mental health disorders in most general population studies [24,26,28,31,40,54,55], as opposed to objective assessment of anxiety/mood disorder treatment counts in one general population study [27], is another possible reason for the lack of consensus among the results of general population studies. Nutsford and colleagues stated in their ecological study that objective assessment of mental health status by measuring anxiety/mood disorder treatment count is a better method for assessing mental health status than the subjective measure of anxiety/mood disorder [27].

4.3. Strengths and Limitations of This Study

In this study, we unraveled the current set of circumstances regarding the relationship between maternal exposure to green space and the odds of antenatal depression during pregnancy after adjusting for a multitude of confounders. Adjustments were done for important maternal confounders, including age, physical activity, and socioeconomic status. Additionally, we accounted for selection bias by including the variable "length of stay at current residence" in the regression models. Our sample size for the main analyses was comparable to the sample sizes used in previous general population studies (442 [28]; 2479 [40]; 4924 [24]; 7552 [27]; 8157 [31]; 24,945 [54]; 65,407 [55] and 260,061 [26]) and one study on women (7547 [13]) that have investigated similar associations between exposure to green space and depression or other mental health outcomes. Our sample size of 6772 women gave sufficient statistical power in determining the association between exposure to green space and antenatal depression, and to generalize our findings to the rest of the pregnant population of New Zealand.

This study had some limitations. Due to the inherent cross-sectional nature of this investigation, causality could not be determined. The CAUs varied significantly in size over the entire sample. More precisely, the CAUs within the region of the Waikato District Health Board were large in size in comparison with the CAUs within the regions of the Auckland and Manukau District Health Boards. Therefore, exposure misclassification for the estimation of green space for large-sized CAUs could not be ruled out. The possibility of green space exposure misclassification for large-sized CAUs was reduced, at least in part, by performing multilevel analysis. Local Councils and LCDB do not capture data on private gardens. Therefore, we could not include private gardens in classification of green spaces in the current study. While it would have been helpful to investigate the role of access to green spaces by determining distances between home addresses and local green spaces, we did not have the data to do so (e.g., the data on home addresses and the data on road/path networks were not available). Visits to, and time spent in green spaces, have been found to be linked to improvements in mental health outcomes. For example, Magdalena van den Berg and colleagues have demonstrated that visits to and time spent in green spaces are associated with better mental health and vitality scores [56]. In our study, we were not determining whether the participants of the *Growing up in New Zealand* study visited and spend their time in green spaces for any reason (e.g., to establish social connections). Also, we did not have data on the quality characteristics of green spaces (e.g., the safety, esthetics, amenities, and level of maintenance of the area). It has been shown that quality of green space is one factor that determines visits to and subsequent use of green space [57].

4.4. Future Directions

Future investigators could involve both objective measurements of access to green spaces by determining distances to local green spaces through the use of roads (people living away from green space often access their local green space through cars) or path networks (people living very close to green space often access their local green space on foot) and the percentage green cover. Private gardens could be included in classification of green spaces. The quality characteristics of the green spaces could also be added as independent variables in the regression models. Gathering information on visits to and time spent in green space is crucial for the determination of the mediators of the association between green space and antenatal depression.

4.5. Comparison between Study Results of Papers Based on the Growing Up in New Zealand Cohort

Two previous pregnant population studies from New Zealand have also utilized the "*Growing Up in New Zealand*" dataset as the data source [33,58]. These studies focused on the associations between: (1) green space and physical activity [33], and (2) green space and pregnancy outcomes of birth weight and gestational age [58]. Despite the lack of association between green space and physical activity, the study of green space and physical activity concluded that exposure to green space could result in better pregnancy health through increased participation in physical activity [33]. The study on green space and pregnancy outcomes of birth weight and gestational age concluded that exposure to green space is not associated with either birth weight or gestational age, based on the cohort as a whole [58]. However, associations for gestational age were found to be significant for specific population subgroups [58]. The current study suggests that exposure to green space is not beneficial for reducing antenatal depression based on either the cohort as a whole, or cohort-specific population subgroups. One reason for the different conclusions could be due to confounding factors, as different sets of independent variables were available for inclusion into the regression models for the different studies. In the study of green space and physical activity, only a few independent variables were used, including age, ethnicity, education, employment status, NZDep2006, and preference for the local lifestyle of the neighborhood [33]. It is possible that some degree of confounding adjustment was not achieved. For example, residual confounding could not have been entirely omitted as we could not control for body mass index due to missing data on body mass index. This meant that green space could have affected physical activity in a positive way, by increasing the odds of participation in physical activity [33]. In contrast, a large range of independent variables were available for the study looking at green space and birth weight and gestational age [58]. These included gestational age, fetus gender, maternal education, employment status, area deprivation, age, self-identified ethnicity, smoking, alcohol consumption during pregnancy, antenatal depression, heart disease or high blood pressure during pregnancy, diabetes mellitus during pregnancy, relationship status with biological father, birth place, parity, lead maternity carer, residential rurality, and time lived in current neighborhood [58]. Similarly, in the current study, a large range of independent variables was available, including maternal age, self-identified ethnicity, relationship status, parity, smoking, alcohol consumption, general health status, physical activity during the first trimester and remainder of pregnancy, education, employment status, residential rurality, area deprivation, and length of stay at current residence. It can therefore be expected that confounding factors were controlled for to a greater extent in the pregnancy outcomes [58] and in the present paper.

It must be noted that in the investigation of green space and pregnancy outcomes [58] subgroup analyses suggested that exposure to green space was associated with gestational age for women with low levels of education. In contrast, due to the lack of interactions between green space and relevant factors, we did not perform subgroup analysis for the investigations of green space and physical activity [33], and green space and antenatal depression. The interaction analysis may also go some way to explaining the differences in outcomes from the three *Growing Up in New Zealand* studies.

5. Conclusions

The results of our study suggest that exposure to green space is not an important driver of better mental health (i.e., less antenatal depression) across the cohort of women in our study. Our result should be considered in the context of other studies by policymakers who are involved in policies on the construction of new residential properties in areas of green space. It remains unclear as to whether the provision of new residential areas in greener environments through urban planning could prove beneficial to the mental health of women by lowering odds of antenatal depression and preventing adverse health outcomes associated with antenatal depression. More studies focusing on pregnant women are needed, in both similar and different social contexts, to determine the associations between green space and antenatal depression, considering both exposure and access to green space.

Acknowledgments: Key support and funding for the *Growing Up in New Zealand* study came from New Zealand's Ministry of Social Development, the Ministry of Health and the University of Auckland (with Auckland UniServices Limited). The Families Commission have contributed the most significant funding and support to the cohort. Other agencies that have also contributed funding include the Ministries of Health, Education, Justice, Science and Innovation, Women's Affairs, and Pacific Island Affairs; the Departments of Labor, and Corrections; Te Puni Kōkiri (Ministry of Māori Affairs); New Zealand Police; Sport and Recreation New Zealand; Housing New Zealand; and the Mental Health Commission. Treasury and the Health Research Council also provided support in the development phase of the study, and the Office of Ethnic Affairs, Statistics New Zealand and the Children's Commission provided consultation.

Author Contributions: The authors Vikram Nichani, Kim Dirks, Bruce Burns and Cameron Grant designed the study. Author Vikram Nichani carried out the literature review, acquired the green space/health data, carried out the statistical analyses, interpreted the statistical results and drafted the manuscript. Authors Kim Dirks, Bruce Burns, and Cameron Grant edited the manuscript and provided input into the interpretation of the results. All authors approved the final version of the paper. Amy Bird and Cameron Grant are part of the *Growing Up in New Zealand* Research Team and provided expertise and oversight with respect to the use of the cohort dataset.

Conflicts of Interest: The authors declare no conflict of interest.

Abbreviations

The following abbreviations are used in this manuscript:

ArcGIS	Aeronautical Reconnaissance Coverage Geographic Information System
CAU	Census Area Unit
CI	Confidence Interval
EPDS	Edinburgh Postnatal Depression Scale
GHQ	General Health Questionnaire
LCDB	Land Cover Database
NZDep2006	New Zealand Deprivation Index 2006
OR	Odds Ratio
SD	Standard Deviation
USA	United States of America

References

1. World Health Organization. Depression. Available online: http://www.who.int/topics/depression/en/ (accessed on 23 November 2016).
2. Greenberg, P.E.; Fournier, A.A.; Sisitsky, T.; Pike, C.T.; Kessler, R.C. The economic burden of adults with major depressive disorder in the United States (2005 and 2010). *J. Clin. Psychiatry* **2015**, *76*, 155–162. [CrossRef] [PubMed]
3. Bromet, E.; Andrade, L.H.; Hwang, I.; Sampson, N.A.; Alonso, J.; de Girolamo, G.; de Graaf, R.; Demyttenaere, K.; Hu, C.; Iwata, N.; et al. Cross-national epidemiology of DSM-IV major depressive episode. *BioMed Cent. Med.* **2011**, *9*, 90. [CrossRef] [PubMed]
4. Andrews, G. Should depression be managed as a chronic disease? *BMJ Br. Med. J.* **2001**, *322*, 419–421. [CrossRef]
5. World Health Organization. *Mental Health—A Call for Action by World Health Ministers*; World Health Organization: Genava, Switzerland, 2001.
6. Leigh, B.; Milgrom, J. Risk factors for antenatal depression, postnatal depression and parenting stress. *BioMed Cent. Psychiatry* **2008**, *8*, 24. [CrossRef] [PubMed]
7. Ministry of Health. *Healthy Beginnings: Developing Perinatal and Infant Mental Health Services in New Zealand*; Ministry of Health: Wellington, New Zealand, 2012.
8. Accortt, E.E.; Cheadle, A.C.D.; Schetter, C.D. Prenatal depression and adverse birth outcomes: An updated systematic review. *Mater. Child Health J.* **2015**, *19*, 1306–1337. [CrossRef] [PubMed]
9. Goedhart, G.; Snijders, A.C.; Hesselink, A.E.; van Poppel, M.N.; Bonsel, G.J.; Vrijkotte, T.G.M. Maternal depressive symptoms in relation to perinatal mortality and morbidity: Results from a large multiethnic cohort study. *Psychosom. Med.* **2010**, *72*, 769–776. [CrossRef] [PubMed]
10. Chaaya, M.; Campbell, O.M.R.; Kak, F.E.; Shaar, D.; Harb, H.; Kaddour, A. Postpartum depression: prevalence and determinants in Lebanon. *Arch. Women Ment. Health* **2002**, *5*, 65–72. [CrossRef] [PubMed]
11. Milgrom, J.; Gemmill, A.W.; Bilszta, J.L.; Hayes, B.; Barnett, B.; Brooks, J.; Ericksen, J.; Ellwood, D.; Buist, A. Antenatal risk factors for postnatal depression: A large prospective study. *J. Affect. Disord.* **2008**, *108*, 147–157. [CrossRef] [PubMed]
12. Stewart, D.E.; Robertson, E.; Dennis, C.L.; Grace, S.L.; Wallington, T. *Postpartum Depression: Literature Review of Risk Factors And Interventions*; University Health Network Women's Health Program: Toronto, ON, USA, 2003.
13. McEachan, R.R.C.; Prady, S.L.; Smith, G.; Fairley, L.; Cabieses, B.; Gidlow, C.; Wright, J.; Dadvand, P.; van Gent, D.; Nieuwenhuijsen, M.J. The association between green space and depressive symptoms in pregnant women: moderating roles of socioeconomic status and physical activity. *J. Epidemiol. Community Health* **2016**, *70*, 253–259. [CrossRef] [PubMed]
14. United States Environmental Protection Agency. What Is Open Space/Green Space? Available online: http://www.epa.gov/region1/eco/uep/openspace.html (accessed on 1 December 2014).
15. Sanders, T.; Feng, X.; Fahey, P.P.; Lonsdale, C.; Astell-Burt, T. The influence of neighborhood green space on children's physical activity and screen time: Findings from the longitudinal study of Australian children. *Int. J. Behav. Nutr. Phys. Act.* **2015**, *12*, 126. [CrossRef] [PubMed]

16. Cetin, M. Determining the bioclimatic comfort in Kastamonu City. *Environ. Monit. Assess.* **2015**, *187*, 640. [CrossRef] [PubMed]
17. Cetin, M. Using GIS analysis to assess urban green space in terms of accessibility: Case study in Kutahya. *Int. J. Sustain. Dev. World Ecol.* **2015**, *22*, 420–424. [CrossRef]
18. Cetin, M.; Sevik, H. Evaluating the recreation potential of IlgazMountain National Park in Turkey. *Environ. Monit. Assess.* **2016**, *188*, 52. [CrossRef] [PubMed]
19. Dadvand, P.; de Nazelle, A.; Figueras, F.; Basagaña, X.; Su, J.; Amoly, E.; Jerrett, M.; Vrijheid, M.; Sunyer, J.; Nieuwenhuijsen, M.J. Green space, health inequality and pregnancy. *Environ. Int.* **2012**, *40*, 110–115. [CrossRef] [PubMed]
20. Markevych, I.; Fuertes, E.; Tiesler, C.M.T.; Birk, M.; Bauer, C.-P.; Koletzko, S.; von Berg, A.; Berdel, D.; Heinrich, J. Surrounding greenness and birth weight: Results from the GINIplus and LISAplus birth cohorts in Munich. *Health Place* **2014**, *26*, 39–46. [CrossRef] [PubMed]
21. Dadvand, P.; Wright, J.; Martinez, D.; Basagaña, X.; McEachan, R.R.C.; Cirach, M.; Gidlow, C.J.; de Hoogh, K.; Gražulevičienė, R.; Nieuwenhuijsen, M.J. Inequality, green spaces, and pregnant women: Roles of ethnicity and individual and neighborhood socioeconomic status. *Environ. Int.* **2014**, *71*, 101–108. [CrossRef] [PubMed]
22. Dadvand, P.; Sunyer, J.; Basagaña, X.; Ballester, F.; Lertxundi, A.; Fernández-Somoano, A.; Estarlich, M.; García-Esteban, R.; Mendez, M.A.; Nieuwenhuijsen, M.J. Surrounding greenness and pregnancy outcomes in four Spanish birth cohorts. *Environ. Health Perspect.* **2012**, *120*, 1481–1487. [CrossRef] [PubMed]
23. Barton, J.; Pretty, J. What is the Best Dose of nature and green exercise for improving mental health? A multi-study analysis. *Environ. Sci. Technol.* **2010**, *44*, 3947–3955. [CrossRef] [PubMed]
24. Bos, E.H.; van der Meulen, L.; Wichers, M.; Jeronimus, B.F. A Primrose Path? Moderating effects of age and gender in the association between green space and mental health. *Int. J. Environ. Res. Public Health* **2016**, *13*, 492. [CrossRef] [PubMed]
25. De Vries, S.; Verheij, R.A.; Groenewegen, P.P.; Spreeuwenberg, P. Natural environments-healthy environments? An exploratory analysis of the relationship between greenspace and health. *Environ. Plan. A* **2003**, *35*, 1717–1731. [CrossRef]
26. Astell-Burt, T.; Feng, X.; Kolt, G.S. Mental health benefits of neighborhood green space are stronger among physically active adults in middle-to-older age: Evidence from 260,061 Australians. *Prev. Med.* **2013**, *57*, 601–606. [CrossRef] [PubMed]
27. Nutsford, D.; Pearson, A.L.; Kingham, S. An ecological study investigating the association between access to urban green space and mental health. *Public Health* **2013**, *127*, 1005–1011. [CrossRef] [PubMed]
28. Nutsford, D.; Pearson, A.L.; Kingham, S.; Reitsma, F. Residential exposure to visible blue space (but not green space) associated with lower psychological distress in a capital city. *Health Place* **2016**, *39*, 70–78. [CrossRef] [PubMed]
29. Morton, S.M.B.; Atatoa Carr, P.E.; Grant, C.C.; Robinson, E.M.; Bandara, D.K.; Bird, A.; Ivory, V.C.; Kingi, T.K.R.; Liang, R.; Marks, E.J.; et al. Cohort profile: Growing up in New Zealand. *Int. J. Epidemiol.* **2013**, *42*, 65–75. [CrossRef] [PubMed]
30. Statistics, New Zealand. Geographic Definitions. Available online: http://www.stats.govt.nz/Census/about-2006-census/2006-census-definitions-questionnaires/definitions/geographic.aspx (accessed on 1 December 2014).
31. Richardson, E.A.; Pearce, J.; Mitchell, R.; Kingham, S. Role of physical activity in the relationship between urban green space and health. *Public Health* **2013**, *127*, 318–324. [CrossRef] [PubMed]
32. Richardson, E.; Pearce, J.; Mitchell, R.; Day, P.; Kingham, S. The association between green space and cause-specific mortality in urban New Zealand: An ecological analysis of green space utility. *BioMed Cent. Public Health* **2010**, *10*, 240. [CrossRef] [PubMed]
33. Nichani, V.; Dirks, K.; Burns, B.; Bird, A.; Morton, S.; Grant, C. Green space and physical activity in pregnant women: Evidence from the growing up in New Zealand study. *J. Phys. Act. Health* **2016**, *13*, 1341–1350. [CrossRef] [PubMed]
34. Auckland Council Park Extent. Available online: http://aucklandopendata.aucklandcouncil.opendata.arcgis.com/datasets/b73e1d6e2fae4515b517db8c975f85c7_0 (accessed on 1 September 2016).
35. Land Cover Database (LCDB). *The New Zealand Land Cover Database (LCDB)*, 3.3 ed.; Land Cover Database (LCDB): Wellington, New Zealand, 2013.

36. Salmond, C.; Crampton, P.; Atkinson, J. *NZDep2006 Index of Deprivation*; University of Otago: Dunedin, New Zealand, 2007.

37. Witten, K.; Hiscock, R.; Pearce, J.; Blakely, T. Neighborhood access to open spaces and the physical activity of residents: A national study. *Prev. Med.* **2008**, *47*, 299–303. [CrossRef] [PubMed]

38. McCormack, G.R.; Shiell, A. In search of causality: A systematic review of the relationship between the built environment and physical activity among adults. *Int. J. Behav. Nutr. Phys. Act.* **2011**, *8*, 125. [CrossRef] [PubMed]

39. Mokhtarian, P.L.; Cao, X. Examining the impacts of residential self-selection on travel behavior: A focus on methodologies. *Transp. Res. Part B Methodol.* **2008**, *42*, 204–228. [CrossRef]

40. Beyer, K.; Kaltenbach, A.; Szabo, A.; Bogar, S.; Nieto, F.; Malecki, K. Exposure to neighborhood green space and mental health: Evidence from the survey of the health of Wisconsin. *Int. J. Environ. Res. Public Health* **2014**, *11*, 3453–3472. [CrossRef] [PubMed]

41. Zhang, Y.; van Dijk, T.; Tang, J.; van den Berg, A.E. Green Space Attachment and Health: A comparative study in two urban neighborhoods. *Int. J. Environ. Res. Public Health* **2015**, *12*, 14342–14363. [CrossRef] [PubMed]

42. Alcock, I.; White, M.P.; Wheeler, B.W.; Fleming, L.E.; Depledge, M.H. Longitudinal effects on mental health of moving to greener and less green urban areas. *Environ. Sci. Technol.* **2014**, *48*, 1247–1255. [CrossRef] [PubMed]

43. Morton, S.; Atatoa Carr, P.E.; Bandara, D.; Grant, C.; Ivory, V.C.; Kingi, T.K.R.; Liang, R.; Perese, L.M.; Pryor, J.E.; Reese, E.; et al. Growing up in New Zealand. In *A Longitudinal Study of New Zealand Children and Their Families. Report 1: Before We Are Born*; Growing Up In New Zealand: Auckland, New Zealand, 2010. Available online: http://www.growingup.co.nz/en.html (accessed on 1 December 2016).

44. Cox, J.L.; Holden, M.; Sagovsky, R. Detection of postnatal depression development of the 10-item Edinburgh Postnatal Depression Scale. *Br. J. Psychiatry* **1987**, *150*, 782–786. [CrossRef] [PubMed]

45. Matthey, S.; Barnett, B.; White, T. The Edinburgh Postnatal Depression Scale. *Br. J. Psychiatry* **2003**, *182*, 368. [CrossRef] [PubMed]

46. Waldie, K.E.; Peterson, E.R.; D'Souza, S.; Underwood, L.; Pryor, J.E.; Carr, P.A.; Grant, C.; Morton, S.M.B. Depression symptoms during pregnancy: Evidence from Growing Up in New Zealand. *J. Affect. Disord.* **2015**, *186*, 66–73. [CrossRef] [PubMed]

47. Tsai, A.C.; Scott, J.A.; Hung, K.J.; Zhu, J.Q.; Matthews, L.T.; Psaros, C.; Tomlinson, M. Reliability and Validity of Instruments for Assessing Perinatal Depression in African Settings: Systematic Review and Meta-Analysis. *PLoS ONE* **2013**, *8*, e82521. [CrossRef] [PubMed]

48. Kheirabadi, G.R.; Maracy, M.R.; Akbaripour, S.; Masaeli, N. Psychometric properties and diagnostic accuracy of the Edinburgh Postnatal Depression Scale in a sample of Iranian women. *Iran. J. Med. Sci.* **2012**, *37*, 32–38. [PubMed]

49. Ekeroma, A.J.; Ikenasio-Thorpe, B.; Weeks, S.; Kokaua, J.; Puniani, K.; Stone, P.; Foliaki, S.A. Validation of the Edinburgh Postnatal Depression Scale (EPDS) as a screening tool for postnatal depression in Samoan and Tongan women living in New Zealand. *N. Z. Med. J.* **2012**, *125*, 41–49. [PubMed]

50. Diez Roux, A.V. A glossary for multilevel analysis. *J. Epidemiol. Community Health* **2002**, *56*, 588–594. [CrossRef] [PubMed]

51. The American College of Obstetricians and Gynecologists (ACOG). The ACOG committee opinion No. 267: Exercise during pregnancy and the postpartum period. *Obstet. Gynecol.* **2002**, *99*, 171–173.

52. Dekrout, A.S.; Clarkson, B.D.; Parsons, S. Temporal and spatial distribution and habitat associations of an urban population of New Zealand long-tailed bats (*Chalinolobus tuberculatus*). *N. Z. J. Zool.* **2014**, *41*, 285–297. [CrossRef]

53. Freeman, C.; Buck, O. Development of an ecological mapping methodology for urban areas in New Zealand. *Landsc. Urban Plan.* **2003**, *63*, 161–173. [CrossRef]

54. Annerstedt, M.; Östergren, P.O.; Björk, J.; Grahn, P.; Skärbäck, E.; Währborg, P. Green qualities in the neighborhood and mental health—Results from a longitudinal cohort study in Southern Sweden. *BioMed Cent. Public Health* **2012**, *12*, 337. [CrossRef] [PubMed]

55. Astell-Burt, T.; Mitchell, R.; Hartig, T. The association between green space and mental health varies across the lifecourse. A longitudinal study. *J. Epidemiol. Community Health* **2014**, *68*, 578–583. [CrossRef] [PubMed]

56. Van den Berg, M.; van Poppel, M.; van Kamp, I.; Andrusaityte, S.; Balseviciene, B.; Cirach, M.; Danileviciute, A.; Ellis, N.; Hurst, G.; Masterson, D.; et al. Visiting green space is associated with mental health and vitality: A cross-sectional study in four European cities. *Health Place* **2016**, *38*, 8–15. [CrossRef] [PubMed]

57. McCormack, G.R.; Rock, M.; Toohey, A.M.; Hignell, D. Characteristics of urban parks associated with park use and physical activity: A review of qualitative research. *Health Place* **2010**, *16*, 712–726. [CrossRef] [PubMed]

58. Nichani, V.; Dirks, K.; Burns, B.; Bird, A.; Morton, S.; Cameron, G. Green space and pregnancy outcomes: Evidence from the growing up in New Zealand study. *Health Place* **2017**, *46*, 21–28. [CrossRef] [PubMed]

International Journal of
*Environmental Research
and Public Health*

MDPI

Article

Does Residential Green and Blue Space Promote Recovery in Psychotic Disorders? A Cross-Sectional Study in the Province of Utrecht, The Netherlands

Susanne Boers [1], **Karin Hagoort** [1,*], **Floortje Scheepers** [1] and **Marco Helbich** [2]

[1] Department of Psychiatry, Faculty of Medicine, University Medical Centre Utrecht, Heidelberglaan 100, 3584 CX Utrecht, The Netherlands; susanne.boers@gmail.com (S.B.); f.e.scheepers-2@umcutrecht.nl (F.S.)
[2] Department of Human Geography and Spatial Planning, Faculty of Geosciences, Utrecht University, Princetonlaan 8a, 3584 CB Utrecht, The Netherlands; m.helbich@uu.nl
* Correspondence: K.Hagoort@umcutrecht.nl; Tel.: +31-88-75-560-25

Received: 29 August 2018; Accepted: 5 October 2018; Published: 8 October 2018

Abstract: Mental health is reportedly influenced by the presence of green and blue space in residential areas, but scientific evidence of a relation to psychotic disorders is scant. We put two hypotheses to the test: first, compared to the general population, psychiatric patients live in neighborhoods with less green and blue space; second, the amount of green and blue space is negatively associated with the duration of hospital admission. The study population consisted of 623 patients with psychotic disorders who had been admitted to the psychiatric ward of an academic hospital in Utrecht, The Netherlands from 2008 to 2016. Recovery was measured by length of stay. Structured patient data was linked to socio-economic status and the amount of green and blue space in the residential area. Associations were assessed by means of regression models controlling for confounding factors. Compared to the general population, psychiatric patients had a significantly lower amount of green space in their neighborhood. This result was not confirmed for blue space. Furthermore, no significant associations were found between green and blue space and the duration of hospital stay. In conclusion, previous studies focusing on other mental disorders, like anxiety or depression, found positive mental health effects of green and blue space in the neighborhood. We were not able to confirm significant effects among our study population on duration of admission, however. Future research focusing on psychotic patients could investigate the influence of exposure to green and blue space on other influences and outcomes on mental health.

Keywords: schizophrenia; psychotic disorders; health data; environmental factors; green space; blue space

---

1. Introduction

A high burden of disease is caused by mental disorders [1], particularly those in the psychotic spectrum, which account for a large share of the disease-adjusted life years attributed to mental disorders [2]. To decrease the burden, we need to understand the factors that influence psychotic disorders and the interaction between these. Besides genetic, biological, psychological, and social factors, which are often associated with psychotic disorders, environmental factors may also be involved [3]. Common aspects of the natural environment near people's homes are green space (a park or woodland) and blue space (bodies of water). Many studies have examined the relation of green and blue space with general mental health [3–7], but research into the effect of green and blue space on specific mental disorders is scarce. In some experimental studies, exposure to green space was found to improve mental health [8–11]. In comparison to the built environment [12], green space is thought to improve social cohesion [13], reduce stress, facilitate restoration of attention problems and fatigue, improve mood [14], and reduce symptoms of major depression [14–16]. Exposure to blue space has been found to reduce stress [17]. Several cross-sectional studies have revealed a negative association between green space, depression [18], suicide mortality [19], and stress [20] but a positive association between time spent in green space and mental health improvement [21]. A recent systematic review [6] described 28 studies that had investigated the possible benefit of green and blue space on mental health. The findings were inconsistent, however. The reviewer cited the limited number of studies that were included and the heterogeneity across the studies regarding the assessment of green and blue space. Thus far, little research has dealt with the influence of green or blue space on psychotic disorders.

This study addresses that gap in the literature by investigating the possible protective effect of living near green or blue space among patients with diagnosed psychotic disorders in the province of Utrecht, The Netherlands. Specifically, it assesses the relation between green space or blue space in the residential environment of patients that have been diagnosed with a psychotic disorder and admitted to an academic hospital. We hypothesized that psychiatric patients live in neighborhoods where the absence of green or blue space is more pronounced than in areas occupied by the general population. Then we tested whether the availability of green or blue space is negatively associated with the duration of admission (as a measure of recovery) for the psychotic disorder.

2. Materials and Methods

2.1. Study Population

In 2013, the Province of Utrecht had 640,610 addresses and 1,245,294 inhabitants [22], of whom 5056 (0.4%) had a diagnosis of 'schizophrenia or any other psychotic disorder' [23]. The study population consisted of patients with a psychotic disorder who had been admitted to the inpatient department of psychiatry, University Medical Center (UMC) Utrecht, between January 2008 and June 2016. After selecting patients whose primary diagnosis was a psychotic disorder, we filtered out those living outside the province based on their residential address. This resulted in the inclusion of 48% of all patients. If one patient had multiple treatments within the study period, we used data from the latest admission because it was up to date. In total, 623 patients were included in the analysis. Approval for the study was obtained from the medical ethics review committee.

2.2. Health Data

The health data originate from the UMC Utrecht clinical databases. To guarantee privacy, data from patients who used the inpatient services of the psychiatric department of the UMC Utrecht were anonymized and stored in the UMC Utrecht research database for analysis. We selected patients with the diagnosis 'schizophrenia spectrum and other psychotic disorders' according to the Diagnostic and Statistical Manual of Mental Disorders (DSM-IV). As dependent variables for subsequent regression models, we operationalized the length of stay as a measure of severity/recovery. The length of stay refers to the number of days for which patients had been admitted to the psychiatric ward at the UMC Utrecht, so the duration could only be calculated for inpatients.

This study was approved by the medical ethics committee of the UMC Utrecht (reference number WAG/nt/16/033895).

2.3. Measures of Green and Blue Space

The key explanatory variables were the availability of green and the availability of blue space, both given in percentages, near the patient's home. Green space refers to agricultural areas, natural areas, or artificially installed greenery; blue space refers to fresh water or salt water bodies. Both green and blue space data were extracted from the most recent Dutch land use database for 2012 with a spatial resolution of 25 × 25 m per raster cell [24]. This land use database differentiates 43 land use types, which were reclassified and aggregated using the ArcGIS 10.4 software (ESRI, Redlands, CA, USA) as follows: green space comprises agricultural areas (categories 1–6, 9), forests (categories 11–12), natural areas (categories 30–43, 45, and 61–62); blue space comprises fresh water and salt water (categories 16 and 17). In accordance with prior studies [6,25], we considered the amount of green and blue space within a circular buffer of 300 m centered on a patient's home address (in %).

2.4. Covariates

Besides environmental exposure, the analyses considered both individual and area-based covariates. Covariates on an individual level were gender and age (in years). Covariates based on area were urbanicity and socio-economic status. Urbanicity adjusts for urban-rural differences in mental health [26] and is operationalized as address density. Addresses were extracted from the Dutch cadaster for the year 2016 and we computed the number of address locations within a 300-m buffer around the patients' home location. Socio-economic status of the neighborhood is represented by means of average residential property value (in 1000 Euros). This variable was collected from Statistics Netherlands on a 100-m grid for the year 2012.

2.5. Statistical Analyses

We used descriptive statistics to summarize the data. To test the hypothesis that psychiatric patients live in neighborhoods with less green and blue space compared to the general population, we performed a series of Chi^2 tests. We carried out non-parametric Spearman correlations between the variables to gain insight into the bivariate associations and to identify problems due to multicollinearity. We selected the Spearman correlation coefficient because our data did not exactly follow a normal distribution. To study the extent to which natural environments (green and blue space) are associated with the severity of symptoms of psychotic patients, we performed a multivariate regression analysis by means of fitting ordinary least squares regression models. We regressed the dependent variable 'length of stay' on blue space and green space while adjusting for the covariates including gender, age, socio-economic status and urbanicity. We estimated two different models with increasing complexity. Model 1 included green and blue space and was adjusted for gender and age. Model 2 additionally considered urbanicity and area-based socio-economic status. SPSS 22.0 (SPSS Inc., Chicago, IL, USA) was used for the data preparation and statistical analysis.

3. Results

3.1. Descriptive Statistics of the Study Population

Our study population consisted of 623 patients, of whom 70% were male (Table 1). The mean age was 38 years, ranging from 11 to 94 years. The patient records included visits between 2008 and 2016 whereas the average length of stay was 40 days and varied widely between 1 and 361 days. About 50% of the patients were admitted for a period shorter than a month. The mean availability of green and blue space in the residential environment is relatively low, and with 5.6% green and 4.2% blue, greenery and water are almost equally available. Both distributions were slightly right-skewed, especially for green space. Approximately 65% of the patients had no green space near their home. The average residential property value was 238,000 Euros. However, the residential property value varied between 90,000 Euros and 1,120,000 Euros. The mean number of addresses was 1085. The standard deviation, at 548, was high.

Table 1. Descriptive statistics of the study population.

Variable	Range	Mean	SD
Length of stay (days)	1–361	40	38.3
Green space (300 m)	0–94	5,6	14.4
Blue space (300 m)	0–46	4,2	5.0
Gender (female, male)	30%; 70%		
Age (years)	11–94	38	14
Socio-economic status (SES) (in 1000 €)	90–1120	228	102
Address density (300 m)	5–2677	1085	548

3.2. Bivariate Analyses

Table 2 shows the bivariate associations between the variables. Green space and blue space were not significantly correlated with the length of stay, however, confirming the expected direction. Older people had a shorter duration of admission. Chi^2 tests showed that psychiatric patients live significantly more often in neighbourhoods with no green space than the general population of the province of Utrecht (Chi^2 = 44.770; *p*-value < 0.010). This result was not confirmed for blue space. The proportion of psychiatric patients with no blue space within their residential area did not significantly differ from the general population (Chi^2 = 0.147; *p*-value = 0.701).

Table 2. Spearman correlation coefficients.

		Green Space	Blue Space	Gender	Age	SES	Urbanicity
Length of stay	Correlation	0.018	0.023	−0.031	−0.200	0.026	−0.010
	p-value	0.653	0.577	0.441	0.000	0.542	0.808
Green space	Correlation	1	−0.125	−0.050	−0.081	0.140	−0.574
	p-value		0.002	0.217	0.045	0.001	0.000
Blue space	Correlation	−0.125	1	0.030	−0.009	0.078	0.004
	p-value	0.002		0.460	0.818	0.070	0.913
Gender	Correlation	−0.050	0.030	1	0.189	0.042	−0.018
	p-value	0.217	0.460		0.000	0.315	0.654
Age	Correlation	−0.081	−0.009	0.189	1	−0.106	0.093
	p-value	0.045	0.818	0.000		0.011	0.019
SES	Correlation	0.140	0.078	0.042	−0.106	1	−0.328
	p-value	0.001	0.070	0.315	0.011		0.000
Urbanicity	Correlation	−0.574	0.004	−0.018	0.093	−0.328	1
	p-value	0.000	0.913	0.654	0.019	0.000	

3.3. Regression

The *F*-test of the partially adjusted regression model 1 (*F* = 4.588; *p* = 0.001) and the fully adjusted model 2 (*F* = 3.057; *p* = 0.006) showed statistical significance (Table 3). Both adjusted R^2 are, at 3%, low. Independent of the fitted model, we found no statistical evidence that either green space or blue space is correlated with our response variable, namely length of stay of patients diagnosed with schizophrenia and other psychotic disorders. Because of a skewed distribution of green and blue space, both variables were also log transformed and grouped. However, these approaches did not change our results; both green and blue space remained insignificant. Similarly, urbanicity and socio-economic status were not related with the response variable. Only age, consistently across both models, was negatively related and statistically significant.

Table 3. Regression results.

	Model	Coef. Unstandardized	Std. Error	Coef. Standardized	*t*-Values	*p*-Values
	Intercept	55.653	4.960		11.220	0.000
	Green space	0.038	0.163	0.010	0.232	0.817
1	Blue space	0.381	0.343	0.047	1.113	0.266
	Male	0.503	3.578	0.006	0.141	0.888
	Age	−0.466	0.118	−0.173	−3.951	0.000
	Intercept	53.002	8.100		6.544	0.000
	Green space	0.059	0.185	0.015	0.321	0.748
	Blue space	0.392	0.346	0.049	1.133	0.258
2	Male	0.546	3.596	0.007	0.152	0.879
	Age	−0.468	0.119	−0.174	−3.934	0.000
	SES	0.005	0.017	0.012	0.278	0.781
	Urbanicity	0.001	0.004	0.018	0.371	0.711

4. Discussion

While the link between natural environment, particularly green space, and health is stressed in numerous studies with various study designs, less is known about the beneficial effects of blue space. Although previous studies focused on green space and mental health [3–7], little research is dedicated to patients with a psychotic disorder and how exposure to both green space and blue space in the residential area would affect people's length of stay in mental health institutions or hospitals.

This is the first study, to our knowledge, investigating the associations between green space and blue space in patients diagnosed with schizophrenia or another psychotic disorder. Our hypothesis that patients with a psychotic disorder reside in neighborhoods with less green and blue space availability compared to the general population was partly confirmed. While we found that psychiatric patients live in areas with less green space, no evidence was found for blue space. Moreover, our regression models did not confirm that green and blue space availability is negatively associated with the duration of hospital admission. Three explanations are plausible to explain the latter result. First, it might be that length of stay in a hospital is a less appropriate outcome measure to unravel the relation between the natural environment and psychotic disorders. One could argue that the natural environment where people live would not be protective during hospital stays. Apparently, according to our results, more green and blue space availability does not result in a shorter admission period. Conceivably, the length of stay does not give a valid indication of the extent of recovery. It should be kept in mind that other studies investigated less severe, self-reported mental health outcomes [26], such as depression symptoms or the quality of life, and not clinical data.

A second possible explanation is that the radius of 300 m that we set for this study was too small, though it proved to be a valid distance to represent the immediate surroundings [6]. Other Dutch studies have drawn a larger radius, for example 1000 or 3000 m. A study assessing the association between green and blue space availability on anxiety, mood and substance use disorder found a significant influence of green space availability in a 1000-meter radius for anxiety disorder but not for any other disorder, nor for blue space availability [26]. A study on the influence of either grass or tree greenness found only positive results at a radius of 1000 m on self-reported health and no effects at 300 m [27].

A third reason could be that the effect of the natural environment on the status of psychotic disorders is less pronounced than it is on other psychiatric disorders. Most studies on mental health and the natural environment have included people with anxiety or mood disorders [6], hypothesizing that the natural environment decreases stress and increases exercise time, thereby reducing symptoms of anxiety or depression.

Several limitations should be considered when interpreting our results. First, though useful for hypothesis generation, the research design is cross-sectional. We recommend that future studies be longitudinal. Second, although we controlled for demographics and socio-economic status, our models remained unadjusted for other factors (e.g., lifestyle, physical activity) known to affect mental health. Third, our sample showed limited variance concerning the availability of green and blue space for inhabitants of Utrecht as well as for patients with schizophrenia and psychotic disorders. Further, the skewness in these two environmental variables made them less powerful as predictors in the analyses. The amounts of green and blue space are calculated around a patient's home address that was registered by the hospital. But as argued elsewhere [3], people spend a limited amount of time at home and are exposed to other environments along their daily trajectories or over the course of their life. Fourth, limited due to data availability, SES was only represented through the proxy variable property value. Other variables such as homeownership may be more suitable.

In view of the mixed findings we suggest three ways to pursue more reliable insight into the interaction between green and blue space availability and psychotic disorders. First, the majority of studies have focused on the natural environment, either in the neighborhood or as operationalized through buffers, rather than using nature exposure during the day (e.g., along walks and during activities or at the workplace). Advances in geotechnology and GPS make it possible to refine exposure assessments in future research. Similarly, historical data on natural environmental exposure has rarely been incorporated in research, although the influence of exposure over the life course might be of great interest [3,28].

5. Conclusions

This study explored how natural environments may be associated with psychotic disorders. Our results did not support the increasingly reported mental health benefit of green and blue space. Across our regression models, the associations tuned out to be insignificant, though the direction of the effect was in line with our hypothesis. Nonetheless, this study does contribute to the literature by exploring the relationship between environments and psychotic disorders by analyzing clinical data, which so far has rarely been attempted.

Author Contributions: Conceptualization, F.S. and M.H.; Formal analysis, S.B., K.H. and M.H.; Investigation, S.B.; Writing—Original Draft, S.B. and K.H.; Writing—Review & Editing, F.S. and M.H.

Funding: Marco Helbich was partly funded from the European Research Council (ERC) under the European Union's Horizon2020 research and innovation program (grant agreement No. 714993).

Acknowledgments: We acknowledge Femke Coenen for her help in clinical data preparation. We also thank Martin Dijst for his comments at an early stage of this research.

Conflicts of Interest: The authors declare no conflict of interest.

References

1. World Health Organization. *Mental Health Action Plan 2013–2020*, 1st ed.; World Health Organization: Geneva, Switzerland, 2013; pp. 1–48, ISBN 978 92 4 150602.
2. Murray, C.J.; Vos, T.; Lozano, R.; Naghavi, M.; Flaxman, A.D.; Michaud, C.; Ezzati, M.; Shibuya, K.; Salomon, J.A.; Abdalla, S.; et al. Disability-adjusted life years (DALYs) for 291 diseases and injuries in 21 regions, 1990–2010: A systematic analysis for the Global Burden of Disease Study 2010. *Lancet* **2012**, *380*, 2197–2223. [CrossRef]
3. Helbich, M. Toward dynamic urban environmental exposure assessments in mental health research. *Environ. Res.* **2018**, *161*, 129–135. [CrossRef] [PubMed]
4. De Vries, S.; Verheij, R.A.; Groenewegen, P.P.; Spreeuwenberg, P. Natural environments—Healthy environments? An exploratory analysis of the relationship between green space and health. *Environ. Plan A* **2003**, *35*, 1717–1731. [CrossRef]
5. Maas, J.; Verheij, R.A.; Groenewegen, P.P.; De Vries, S.; Spreeuwenberg, P. Green space, urbanity, and health: How strong is the relation? *J. Epidemiol. Community Health* **2006**, *60*, 587–592. [CrossRef] [PubMed]
6. Gascon, M.; Triguero-Mas, M.; Martinez, D.; Dadvand, P.; Forns, J.; Plasència, A.; Nieuwenhuijsen, M.J. Mental health benefits of long-term exposure to residential green and blue spaces: A systematic review. *Int. J. Environ. Res. Public Health* **2015**, *12*, 4354–4379. [CrossRef] [PubMed]

7. Van den Bosch, M.; Sang, Å. Urban natural environments as nature-based solutions for improved public health—A systematic review of reviews. *Environ. Res.* **2017**, *158*, 373–384. [CrossRef] [PubMed]

8. Van den Berg, M.; Wendel-Vos, W.; Van Poppel, M.; Kemper, H.; Van Mechelen, W.; Maas, J. Health benefits of green spaces in the living environment: A systematic review of epidemiological studies. *Urban For. Urban Green.* **2015**, *14*, 806–816. [CrossRef]

9. Sugiyama, T.; Leslie, E.; Giles-Corti, B.; Owen, N. Associations of neighbourhood greenness with physical and mental health: Do walking, social coherence and local social interaction explain the relationships? *J. Epidemiol. Community Health* **2008**, *62*, 9–17. [CrossRef]

10. Triguero-Mas, M.; Dadvand, P.; Cirach, M.; Martínez, D.; Medina, A.; Mompart, A.; Basagaña, X.; Gražulevičienė, R.; Nieuwenhuijsen, M.J. Natural outdoor environments and mental and physical health: Relationships and mechanisms. *Environ. Int.* **2015**, *77*, 35–41. [CrossRef] [PubMed]

11. Maas, J.; Verheij, R.A.; De Vries, S.; Spreeuwenberg, P.; Schellevis, F.G.; Groenewegen, P.P. Morbidity is related to a green living environment. *J. Epidemiol. Community Health* **2009**, *63*, 967–973. [CrossRef] [PubMed]

12. Gong, Y.; Palmer, S.; Gallacher, J.; Marsden, T.; Fone, D. A systematic review of the relationship between objective measurements of the urban environment and psychological distress. *Environ. Int.* **2016**, *96*, 48–57. [CrossRef] [PubMed]

13. Kaczynski, A.T.; Henderson, K.A. Environmental correlates of physical activity: A review of evidence about parks and recreation. *Leis. Sci.* **2007**, *29*, 315–354. [CrossRef]

14. Hartig, T.; Mitchell, R.; De Vries, S.; Frumkin, H. Nature and health. *Annu. Rev. Public Health* **2014**, *35*, 207–208. [CrossRef] [PubMed]

15. Berman, M.G.; Kross, E.; Krpan, K.M.; Askren, M.K.; Burson, A.; Deldin, P.J.; Kaplan, S.; Sherdell, L.; Gotlib, I.H.; Jonides, J. Interacting with nature improves cognition and affect for individuals with depression. *J. Affect Disord.* **2012**, *140*, 300–305. [CrossRef] [PubMed]

16. Kaplan, S. The restorative benefits of nature: Toward an integrative framework. *J. Environ. Psychol.* **1995**, *15*, 169–182. [CrossRef]

17. Völker, S.; Kistemann, T. The impact of blue space on human health and well-being—Salutogenetic health effects of inland surface waters: A review. *Int. J. Hyg. Environ Health* **2011**, *214*, 449–460. [CrossRef] [PubMed]

18. Helbich, M.; Klein, N.; Roberts, H.; Hagedoorn, P.; Groenewegen, P. More green space is related to less antidepressant prescription rates in the Netherlands: A Bayesian geoadditive quantile regression approach. *Environ. Res.* **2018**, *166*, 290–297. [CrossRef] [PubMed]

19. Helbich, M.; de Beurs, D.; Kwan, M.; O'Connor, R.; Groenewegen, P. Natural environments and suicide mortality in the Netherlands—A cross-sectional, ecological study. *Lancet Planet. Health* **2018**, *2*, e134–e139. [CrossRef]

20. Stigsdotter, U.A.; Ekholm, O.; Schipperijn, J.; Toftager, M.; Kamper-Jørgensen, F.; Randrup, T.B. Health promoting outdoor environments—Associations between green space, and health, health-related quality of life and stress based on a Danish national representative survey. *Scand. J. Public Health* **2010**, *38*, 411–417. [CrossRef] [PubMed]

21. Van den Berg, M.; Van Poppel, M.; Van Kamp, I.; Andrusaityte, S.; Balseviciene, B.; Cirach, M.; Danileviciute, A.; Ellis, N.; Hurst, G.; Masterson, D.; et al. Visiting green space is associated with mental health and vitality: A crosssectional study in four European cities. *Health Place* **2016**, *38*, 8–15. [CrossRef] [PubMed]

22. Regionale Kerncijfers Nederland, Totale Bevolking Utrecht (PV). Available online: http://statline.cbs.nl/Statweb/publication/?DM=SLNL&PA=70072ned&D1=0&D2=11&D3=18&HDR=T&STB=G1,G2&VW=T (accessed on 21 January 2017).

23. DBC-Gefinancierde GGZ; Personen per Diagnose, Regio Utrecht (PV). Available online: http://statline.cbs.nl/Statweb/publication/?DM=SLNL&PA=83711ned&D1=a&D2=a&D3=1&D4=16&D5=11&D6=l&HDR=G2,G5,G3,T,G1&STB=G4&VW=T (accessed on 21 January 2017).

24. Hazeu, G.W. Operational land cover and land use mapping in the Netherlands. In *Land Use and Land Cover Mapping in Europe*, 1st ed.; Manakos, I., Braun, M., Eds.; Springer: Dordrecht, The Netherlands, 2014; Volume 18, pp. 283–296, ISBN 978-94-007-7968-6.

25. Peen, J.; Schoevers, R.A.; Beekman, A.T.; Dekker, J. The current status of urban-rural differences in psychiatric disorders. *Acta Psychiatr. Scand.* **2010**, *121*, 84–93. [CrossRef] [PubMed]

26. De Vries, S.; Ten Have, M.; Van Dorsselaer, S.; Van Wezep, M.; Hermans, T.; De Graaf, R. Local availability of green and blue space and prevalence of common mental disorders in the Netherlands. *Br. J. Psychiatry Open* **2016**, *2*, 366–372. [CrossRef] [PubMed]
27. Orban, E.; Sutcliffe, R.; Dragano, N.; Jöckel, K.H.; Moebus, S. Residential surrounding greenness, self-rated health and interrelations with aspects of neighborhood environment and social relations. *J. Urban Health* **2017**, *94*, 158–169. [CrossRef] [PubMed]
28. Pearce, J.; Shortt, N.; Rind, E.; Mitchell, R. Life course, green space and health: Incorporating place into life course epidemiology. *Int. J. Environ. Res. Public Health* **2016**, *13*, 331. [CrossRef] [PubMed]

International Journal of
*Environmental Research
and Public Health*

MDPI

Article

Longitudinal Impact of Hurricane Sandy Exposure on Mental Health Symptoms

Rebecca M. Schwartz [1] [iD], Christina N. Gillezeau [2] [iD], Bian Liu [2], Wil Lieberman-Cribbin [2] [iD]
and Emanuela Taioli [2,*]

[1] Department of Occupational Medicine, Epidemiology and Prevention, Hofstra Northwell Health School of
    Medicine, Great Neck, NY 11021, USA; Rschwartz3@northwell.edu
[2] Department of Population Health Science and Policy and Institute for Translational Epidemiology,
    Icahn School of Medicine at Mount Sinai, New York, NY 10029, USA; cnanderson89@gmail.com (C.N.G.);
    bian.liu@mountsinai.org (B.L.); wil.lieberman-cribbin@icahn.mssm.edu (W.L.-C.)
*   Correspondence: emanuela.taioli@mountsinai.org; Tel.: +1-212-659-9590

Received: 27 July 2017; Accepted: 22 August 2017; Published: 24 August 2017

Abstract: Hurricane Sandy hit the eastern coast of the United States in October 2012, causing billions
of dollars in damage and acute physical and mental health problems. The long-term mental health
consequences of the storm and their predictors have not been studied. New York City and Long Island
residents completed questionnaires regarding their initial Hurricane Sandy exposure and mental
health symptoms at baseline and 1 year later (N = 130). There were statistically significant decreases
in anxiety scores (mean difference = -0.33, $p < 0.01$) and post-traumatic stress disorder (PTSD) scores
(mean difference = -1.98, $p = 0.001$) between baseline and follow-up. Experiencing a combination of
personal and property damage was positively associated with long-term PTSD symptoms (OR_{adj} 1.2,
95% CI [1.1–1.4]) but not with anxiety or depression. Having anxiety, depression, or PTSD at baseline
was a significant predictor of persistent anxiety (OR_{adj} 2.8 95% CI [1.1–6.8], depression (OR_{adj} 7.4
95% CI [2.3–24.1]) and PTSD (OR_{adj} 4.1 95% CI [1.1–14.6]) at follow-up. Exposure to Hurricane Sandy
has an impact on PTSD symptoms that persists over time. Given the likelihood of more frequent
and intense hurricanes due to climate change, future hurricane recovery efforts must consider the
long-term effects of hurricane exposure on mental health, especially on PTSD, when providing
appropriate assistance and treatment.

Keywords: natural disasters; follow-up; anxiety; depression; post-traumatic stress disorder

1. Introduction

On 29 October 2012, Hurricane Sandy hit the Eastern Seaboard causing an estimated 71 billion
dollars in damage [1], displacing 20,000 individuals from their homes, and directly causing the deaths
of 117 people [2]. Additionally, Hurricane Sandy created a host of new hazards and challenges for
those returning to their homes including limited access to gasoline, difficulty traveling due to the
closure of and damage to the New York City subway system [3], and increased exposure to mold
and other environmental toxins [4]. It is generally accepted that increased exposure to traumas are
associated with increased stress and negative mental health outcomes, particularly post-traumatic
stress disorder (PTSD) [5–8], although the social, environmental, and economic factors that exacerbate
or mitigate these outcomes are not well understood. Additionally, most research has focused on the
immediate aftermath of the disaster, while few studies have examined the long-term impact on mental
health outcomes.

Research on Hurricane Katrina indicated that those with increased levels of exposure experienced
worse mental and physical health at least one year after the disaster [9]. Although the number of
stressors decreased over time among those exposed to Katrina, the psychological effects of those

stressors persistently played a prominent role in people's lives [10,11]. Results from Katrina cannot be generalized to all disasters, or even all hurricanes; although survivors of Hurricane Katrina were more likely to develop PTSD symptoms as a result of long-term property damage five years after the storm, as compared to Hurricane Andrew where PTSD symptoms among survivors were more likely to be associated with immediate, not long-term, hurricane exposures [12].

Certain populations are believed to be at increased risk of negative mental health symptoms after a disaster. After Hurricane Sandy, older adults were reported to be especially vulnerable, particularly if they lacked social support [13,14]. Additionally, research suggests that women and individuals who have a history of trauma, including experience coping with previous hurricanes and 9/11, may be at an increased risk of PTSD symptoms [6,15,16].

In recognition of the fact that disasters can create or exacerbate mental health issues, the Federal Emergency Management Agency (FEMA) allocated $50 million toward mental health services after Hurricane Sandy, although a 2015 study reported that 5.9% of exposed individuals still had unmet mental health needs [17]. A more targeted approach, however, is to connect individuals with mental health needs with additional resources, such as through Project Hope [18], although more research is required to predict and understand areas of future mental health needs.

Previous research has already established that increases in mental health disorders after Hurricane Sandy were clustered in geographic areas that were more exposed to the effects of the hurricane, and that within these clusters, minorities were at greater risk of PTSD symptoms [19,20]. However, despite the increase in mental health symptoms, there was a decrease in mental health-related visits to local emergency rooms in the month after Hurricane Sandy [21]. A 2015 study by Schwartz et al. [22] examined the relationship between hurricane exposures and level of perceived stress after Hurricane Sandy. This research reported that perceived stress was higher in areas more strongly affected by Hurricane Sandy, and that individual exposure was significantly associated with increased perceived stress levels [22]. In addition, our previous research indicated that increased exposure was associated with increased PTSD, depression, and anxiety symptoms approximately 1–2 years after the hurricane [23]. Building upon our previous studies, the current work assessed the long-term mental health status of a subgroup of respondents who completed mental health surveys at a second time point, one year after their first survey completion. We hypothesized that mental health symptoms persist over time and that initial Hurricane Sandy exposure will still be associated with mental health symptoms at a follow-up point.

2. Materials and Methods

This was a longitudinal study of mental health symptoms among residents of Nassau, Suffolk, Queens, and Richmond (Staten Island) counties in New York, which has been described previously [22–24]. Initial surveys were conducted 11 to 28 months after Hurricane Sandy at recruitment venues identified in conjunction with community and government partners. Participants were recruited using convenience-sampling techniques from community sites, including senior centers, libraries, gymnasiums, faith-based centers, community colleges, and community centers in both heavily and less affected areas across the region.

Out of the original 673 participants from the baseline survey, 130 subjects (19.3%) participated in the follow-up study, which relied on convenience-sampling techniques via phone, mail, and email recruitment from among the initial participants. One-year follow-up participants were reimbursed for their time with a $20 gift card.

Follow-up surveys were distributed via mail, telephone and email. Participants in the baseline study were called using the phone numbers provided during the initial survey. Voicemail messages were left for individuals who did not answer their phones, and persistent efforts were made to call each individual at different times of the day. Repeated emails were sent to all subjects who had provided their address and had not yet completed the research survey between February and April of 2016; five blasts were sent for each subject. The initial survey was self-administered, while the follow-up survey was self-administered when collected via mail or email, but was researcher-administered when collected via telephone. A small percentage of surveys were administered in person at a local retirement home. Of the 130 surveys, 12 (9.1%) were answered in person, 40 (30.8%) were answered via telephone, 48 (37.0%) were answered via postal mail, and 30 (23.1%) were answered via email.

2.1. Mental Health and Behavioral Outcomes

The primary outcomes measured were mental health symptoms of anxiety, depression and PTSD. Anxiety and depression symptoms were assessed using the previously validated Patient Health Questionnaire-4 (PHQ-4) [25]. Participants were categorized as having anxiety or depression symptoms if they scored a 2 or greater on questions relating to anxiety or depression. PTSD symptoms were assessed using the previously validated Civilian PTSD Questionnaire—Hurricane Sandy Specific (PTSD/PCL$^-$S) [26]. Patients with a score of 30 or greater were categorized as having PTSD symptoms. The Cronbach's alpha was 0.88 for baseline anxiety, 0.81 for follow-up anxiety, 0.86 for baseline depression, 0.78 for follow-up depression, 0.96 for baseline PTSD, and 0.95 for follow-up PTSD. Smoking and alcohol use were also asked about, and were secondary outcomes in the current study. Participants were categorized as smokers if they currently used tobacco and as problem drinkers if they exceeded the *National Institute on Alcohol Abuse and Alcoholism* (NIAAA) assessment guidelines ($\geq$14 drinks/week in men; $\geq$7 women; $\geq$5 drinks on a single occasion in the past week in men; $\geq$4 women) [27].

2.2. Hurricane Exposure

Participants were asked at baseline to answer questions regarding their level of hurricane exposure and these questions were separated into three categories, consistent with the category classifications in Schwartz et al. [22]. Personal exposures dealt with exposures that directly affected the participant or their family, and property exposures related to the level of personal property affected and financial hardship experienced. A total of 16 personal exposures and 14 property exposures were measured; the level of exposure in each category was determined by taking the sum of the number of each kind of exposures an individual experienced, for a total possible score of 30 exposures. Participant total scores ranged between 0 and 21 (Table S1). The Cronbach's alpha was 0.62 for baseline personal exposure and 0.85 for baseline property exposure. A reliability study showed that self-reported total exposure did not change significantly between baseline and one-year interviews (mean difference = 0.13, $p = 0.34$), although baseline exposure scores were used in analyses.

2.3. Covariates

Demographic information including age, sex, race, and education was also collected in the initial survey. Self-reported, physician-diagnosed mental health diagnoses (anxiety disorder, depression, PTSD, schizophrenia, bipolar, substance/alcohol abuse, substance/prescription abuse, or other mental health problems) were recorded during both the first and second survey waves.

2.4. Statistical Analysis

Mental health and behavioral outcomes were treated as dichotomized variables using commonly employed clinical-relevant cutoffs in the main analysis to increase the interpretability of the results, while both dichotomized and continuous mental health outcomes were considered in explorative analyses. The prevalence of mental health symptoms at baseline and follow-up was analyzed using McNemar's Test for dichotomous outcomes and Wilcoxon signed-rank tests for continuous outcomes due to non-normal distribution of these variables (Shapiro-Wilk test for baseline anxiety ($p < 0.0001$); follow-up anxiety ($p < 0.0001$); baseline depression ($p < 0.0001$); follow-up depression ($p < 0.0001$); baseline PTSD ($p < 0.0001$); follow-up PTSD ($p < 0.0001$). Determinants of mental health symptoms at follow-up were assessed using logistic and linear regression models adjusted for the following covariates: age, gender, ethnicity, education, presence of anxiety, depression, or PTSD symptoms at baseline, pre-existing physician-diagnosed mental health condition, time since Hurricane Sandy, and medical insurance status. As the main purpose of the study is to identify the longitudinal impact of hurricane exposure, the analysis was conducted on the 130 participants who had both baseline and follow-up data, excluding the 543 participants who only had baseline data. Chi-square and Wilcoxon-Mann-Whitney tests were used to test for differences between the 130 participants who completed the follow-up and the remaining 543. The majority of the variables were without missing data, with five variables having one to two missing data, which were treated as missing at random. All analyses were performed in SAS version 9.4 (SAS Institute Inc., Cary, NC, USA) and R Studio version 3.2.2 (R Foundation for Statistical Computing, Vienna, Austria).

3. Results

3.1. Characteristics of the Study Population

The majority of the 130 respondents were female (77.7%), white (57.0%), and had at least a high school diploma (94.6%) at baseline, while 65 participants (50%) met the criteria for anxiety, 46 (35.4%) for depression, and 37 (28.7%) for Hurricane Sandy-related PTSD symptoms (Table 1). Most subjects (n = 94; 72.3%) did not report having a history of diagnosed mental health disorders at baseline. Among those who reported a mental health diagnosis, 17 (13.1%) reported a diagnosis prior to, but not after Hurricane Sandy, 8 (6.2%) after Hurricane Sandy, and 11 (8.5%) both before and after the hurricane. On average (mean ± standard deviation), participants had 1.2 ± 1.64 personal exposure scores, 4.04 ± 3.41 property exposure scores and 5.28 ± 4.59 total exposure scores.

Table 1. Characteristics of the study population at baseline.

Variable		N	%
Sex	Male	29	22.31
	Female	101	77.69
Race/Ethnicity (missing n = 2)	White	73	57.03
	Black	25	19.53
	Hispanic Ethnicity	18	14.06
	Other/Mixed	12	9.38
Education (missing n = 1)	<High School	7	5.43
	≥High School	122	94.57
Medical Insurance (missing n = 1)	No	11	8.53
	Yes	118	91.47
Problem Alcohol Drinkers	No	98	75.38
	Yes	32	24.62
Current Smoker	No	116	89.23
	Yes	14	10.77
Mental health History	No	94	72.31
	Yes	36	27.69
Anxiety (PHQ4 score)	<2	65	50.00
	≥2	65	50.00
Depression (PHQ4 score)	<2	84	64.62
	≥2	46	35.38
Post-Traumatic Stress Disorder (PTSD) (PCL-S) (missing n = 1)	<30	92	71.32
	≥30	37	28.68
Mental Health problems	Never	94	72.31
	Before	17	13.08
	After	8	6.15
	Before and After	11	8.46

Table 1. *Cont.*

Variable		N	%
	Mean ± SD	Median	Range
Age (years, missing n = 1)	49.73 ± 20.82	52	18–92
Sex	Male	29	22.31
	Female	101	77.69
Elapsed time between Sandy and Baseline questionnaire (months, missing n = 0)	14.47 ± 3.26	13.13	11.8–27.9
Personal damage score (missing n = 0)	1.25 ± 1.64	1	0–8
Property damage score (missing n = 0)	4.04 ± 3.41	3	0–14
Personal and propertydamage score (missing n = 0)	5.28 ± 4.59	3.5	0–21

The 130 participants who completed the follow-up study were partially representative of the 543 subjects who did not agree to answer follow-up questions about race/ethnicity (χ^2 (3) = 3.48, p = 0.32), education status (χ^2 (1) = 0.44, p = 0.51), medical insurance (χ^2 (1) = 0.28, p = 0.60), and age (p = 0.17). The follow-up sample, however, was composed of a higher proportion of females (79% vs. 61%; χ^2 (1) = 13.7, p < 0.001) and had higher total exposure (5.28 ± 4.59 vs. 3.56 ± 3.76, p < 0.001) compared to the original sample.

3.2. Changes in Mental Health Symptoms between Baseline and Follow-Up

The prevalence of participants with anxiety (50.0% to 41.5%), depression (35.4% to 30.8%), and PTSD (29.2% to 24.8%) symptoms decreased from baseline to follow-up (Table 2). There were statistically significant decreases in anxiety scores (mean difference = −0.33, p < 0.01) and PTSD scores (mean difference = −1.98, p = 0.001) between baseline and follow-up, but differences in depression scores were not statistically significant (mean difference = −0.12, p < 0.39). There were statistically significant correlations between mental health variables at baseline and at follow-up (Table S2). There was no significant association between demographic variables and changes in anxiety, depression, and PTSD scores between baseline and follow-up (data not shown).

Table 2. Prevalence of mental health symptoms at baseline and follow-up.

Variables		Baseline		Follow-Up		*p*-Value
		N	%	N	%	
Anxiety	<2	65	50.00	76	58.5	0.11[a]
	≥2	65	50.00	54	41.5	
	Mean ± SD	129	1.82 ± 1.84	129	1.49 ± 1.73	0.01[b]
Depression	<2	84	64.62	90	69.23	0.38[a]
	≥2	46	35.38	40	30.77	
	Mean ± SD	130	1.22 ± 1.72	130	1.10 ± 1.61	0.39[b]
PTSD	<30	92	71.32	97	75.19	0.44[a]
	≥30	37	58.68	32	24.81	
	Mean ± SD	129	26.95 ± 12.72	129	24.97 ± 11.57	0.001[b]

[a] McNemar's Test; [b] Wilcoxon Signed-Rank Test.

3.3. Factors Associated with Mental Health Symptoms at Follow-Up

Experiencing personal damage (OR_{adj} 1.6, 95% CI [1.2–2.2]), property damage (OR_{adj} 1.3, 95% CI [1.1–1.5]), or a combination of personal and property damage (OR_{adj} 1.2, 95% CI [1.1–1.4]) was positively associated with PTSD symptoms, but not with anxiety or depression symptoms at follow-up (Figure 1). Having anxiety, depression, or PTSD at baseline was a strong predictor of having anxiety (OR_{adj} 2.8, 95% CI [1.1–6.8]), depression (OR_{adj} 7.4, 95% CI [2.3–24.1]) and PTSD (OR_{adj} 4.1, 95% CI [1.1–14.6]) symptoms at follow up. Female gender was positively associated with anxiety (OR_{adj} 3.4, 95% CI [1.1–10.9]), while having a history of mental health concerns prior to Hurricane Sandy was positively associated with PTSD symptoms (OR_{adj} 3.5, 95% CI [1.1–11.0]). When continuous measures of mental health outcomes were used in the models, total exposure was significantly associated with PTSD ($\beta = 0.69$, $p = 0.001$), but not anxiety ($\beta = 0.01$, $p = 0.76$) or depression ($\beta = -0.02$, $p = 0.62$) symptoms (Table 3).

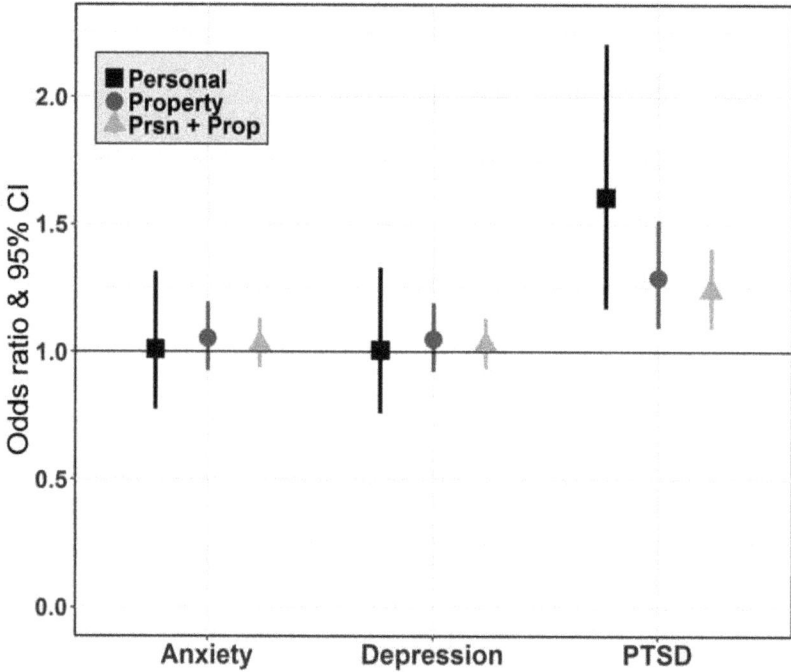

Figure 1. Association between hurricane exposure (personal, property, personal + property) and mental health at follow-up (anxiety, depression, PTSD symptoms). Models were adjusted for age, gender, race, education, medical insurance, existing mental health conditions, elapsed time between Hurricane Sandy and baseline, and mental health condition at baseline.

Table 3. Association between exposure and mental health symptoms.

	Anxiety		Depression		PTSD	
	Estimate	*p*-Value	Estimate	*p*-Value	Estimate	*p*-Value
Total Exposure Model						
Intercept	−0.80722	0.6603	0.79931	0.638	13.67105	0.2423
Mental Health Baseline	1.08977	0.0005	0.88404	0.0021	4.37012	0.0257
Gender	0.62258	0.0903	0.06906	0.8377	0.13119	0.9551
Age	−0.00334	0.6791	−0.00234	0.7529	0.05646	0.2685
Black	−0.48665	0.2135	0.21294	0.5538	4.84212	0.0513
Hispanic	0.03364	0.8798	−0.0386	0.8512	−0.35384	0.8027
Other/Mixed	0.11276	0.8265	1.24303	0.01	0.60231	0.8592
Education	0.11743	0.8531	−0.67738	0.2495	−6.26434	0.1211

Table 3. *Cont.*

	Anxiety		Depression		PTSD	
	Estimate	p-Value	Estimate	p-Value	Estimate	p-Value
Mental Health History	0.90104	0.0093	0.90403	0.0049	6.97867	0.0016
Medical Insurance	0.23852	0.6571	0.09397	0.85	0.06583	0.9852
Elapsed Time	0.01763	0.5895	0.02444	0.419	0.35813	0.0887
Total Exposure	0.00999	0.7626	−0.01537	0.6152	0.69631	0.0012
Personal Exposure Model						
Intercept	−0.66597	0.7154	0.79232	0.6389	15.93477	0.1785
Mental Health Baseline	1.10431	0.0004	0.87524	0.0022	4.94715	0.0126
Gender	0.65571	0.0736	0.07207	0.8299	0.56356	0.8111
Age	−0.00349	0.6658	−0.00286	0.7002	0.07215	0.1661
Black	−0.54218	0.1628	0.20651	0.5625	4.26012	0.0886
Hispanic	0.04301	0.8477	−0.01686	0.9351	−0.95971	0.5088
Other/Mixed	0.10969	0.8311	1.22454	0.0111	1.40813	0.683
Education	0.0987	0.8761	−0.66869	0.2542	−6.85456	0.095
Mental Health History	0.92534	0.0075	0.90824	0.0046	7.1842	0.0014
Medical Insurance	0.21566	0.6891	0.06753	0.8922	0.46097	0.8988
Elapsed Time	0.01663	0.611	0.02378	0.4316	0.37249	0.0819
Personal Exposure	−0.03755	0.6804	−0.05971	0.4779	1.54794	0.0094
Property Exposure Model						
Intercept	−0.88509	0.6296	0.76053	0.6547	13.87039	0.2381
Mental Health Baseline	1.07513	0.0006	0.8802	0.0023	4.28154	0.0301
Gender	0.60858	0.097	0.06018	0.8583	0.23397	0.9204
Age	−0.00365	0.6513	−0.00226	0.7618	0.04895	0.3407
Black	−0.464	0.2334	0.22806	0.5252	4.64593	0.0619
Hispanic	0.0457	0.8378	−0.04226	0.8382	−0.04199	0.9766
Other/Mixed	0.09979	0.8461	1.24465	0.01	0.33993	0.9208
Education	0.13396	0.8326	−0.67249	0.2536	−6.1866	0.1279
Mental Health History	0.89183	0.0099	0.89709	0.0052	7.08582	0.0014
Medical Insurance	0.2298	0.6685	0.10165	0.838	−0.33084	0.9261
Elapsed Time	0.01762	0.5891	0.02474	0.4135	0.347	0.1006
Property Exposure	0.02684	0.5448	−0.01349	0.7423	0.88476	0.0023

Models were adjusted for age, gender, race, education, medical insurance, existing mental health conditions, elapsed time between Hurricane Sandy and baseline, and mental health condition at baseline.

Total exposure to Hurricane Sandy was not statistically associated with smoking (OR_{adj} 1.0, 95% CI [0.9–1.2]) or problem drinking (OR_{adj} 0.9, 95% CI [0.9–1.0]) after adjusting for age, gender, race, education, medical insurance, existing mental health conditions, elapsed time between Hurricane Sandy and baseline, and mental health condition at baseline.

4. Discussion

The current analysis indicates that property and personal hurricane exposures were associated with statistically significant increased odds of PTSD symptoms at follow-up, but not with anxiety or depression symptoms, suggesting that there may be a persistence of the negative impact of hurricanes on PTSD specifically. Previous studies following Hurricane Katrina and Hurricane Sandy have also noted that those exposed to greater stressors and property damage were more likely to demonstrate symptoms of PTSD and mental illness [9,28,29]. Similarly, there is evidence in the literature that the relationship between property exposure and PTSD can also be driven by secondary traumas that are associated with a loss of community and with searching for a new home [12,30]. This has been shown to have been more devastating among low-income individuals after Katrina [12] and Sandy [29], and is possible in this study as well, although we did not see differences by education or insurance status. Only one study has reported on post-traumatic stress at two time points following Sandy [29], and found that disaster-related stressors were significantly associated with higher post-traumatic stress at the first time point [29]. This work, however, sampled a cross-sectional, different set of participants at each time point, and thus was not longitudinal. Additionally, the current work relies on a more comprehensive definition of hurricane exposure, adding financial loss, access to medications, and vehicle loss, which are important determinants of mental health. As such, this work is the first to our knowledge to study longitudinal mental health symptoms within the same sample over time following Hurricane Sandy.

Previous cross-sectional and longitudinal studies of mental health outcomes after Katrina have reported that while a lack of social support does not strictly cause post-traumatic symptoms, a lack of social resources can increase the risk of stressors following a disaster, which in turn, influences post-traumatic stress [10,31–34].

These findings can also be placed into context with the Conservation of Resource theory (COR) [35,36], a framework to understand traumatic stress that has been applied following disasters [15,37–40]. Briefly, COR asserts that "individuals strive to obtain, retain, foster, and protect those things they centrally value", termed resources, which universally include health, well-being, family, and also objects (cars, homes), conditions (employment, marriage), personal traits (self-efficacy, self-esteem), and energy (knowledge, money) [41]. Natural disasters subsequently threaten and destroy these resources and thus create stress. However, those with greater resources are less vulnerable to resource loss and can rebound more robustly compared to those with fewer resources that have less capacity to bounce back [41]. After a natural disaster, disparities exist between those with differing amounts of personal, social and material resources, informing long-term recovery after initial exposure to a natural disaster.

COR has emerged in analyses following Hurricane Katrina [42–44] and Hurricane Sandy [15,45], linking a loss of personal resources to the stress of displacement [42] and the loss of personal property, social support and physical health to long-term psychological distress [43]. However, the presence of personal and social resources can mitigate negative outcomes following exposure [45], but there are differences among people in their capacity to protect and regain resources that help mitigate the stress of hurricane exposure [15]. Although COR was not explicitly tested in this study, this work assessed object resources (property and personal exposure assessments), personal traits (personal exposure assessment) and energy resources (education status) and realized the association between larger initial personal and property exposures and the presence of long-term PTSD symptoms.

Presenting with symptoms of anxiety, depression, or PTSD at baseline were the strongest predictors of having anxiety, depression, or PTSD symptoms at follow-up across all three exposure types (personal, property, personal and property). Additionally, having a history of mental health concerns was a strong predictor of PTSD symptoms. This supports the literature stating that people with pre-existing mental health conditions and exposure to trauma were more vulnerable to these following Hurricane Katrina [46] and Hurricane Sandy [47]. Although average scores for anxiety and PTSD symptoms decreased between the initial and follow-up surveys, this does not reflect universal decreases in mental health symptoms for all participants, nor does it minimize the importance of the high prevalence of anxiety, depression, and PTSD symptoms at follow-up. As such, the results indicate the persistence of poor mental health outcomes in this sub-population exposed to Hurricane Sandy, and the need for future research to integrate measures of social support into mental health assessments in order to mitigate the long-term impacts of hurricane exposure.

There are several limitations to this study. This research relies on self-reported hurricane exposure items, which allows for recall bias, although it should be noted that no significant differences were found in the hurricane exposure assessment between baseline and follow-up, thus strengthening the validity of the exposure assessment. Furthermore, it is possible that people who were heavily affected by the hurricane or who were experiencing mental health difficulties were more prone to participate in research that could potentially mitigate their mental health symptoms, thereby potentially increasing the prevalence of mental health symptoms in the cohort. However, efforts were made to recruit participants from communities throughout the region, including those not highly affected by the hurricane, and from community events and organizations that were not in any way specific to either mental health service provision or Hurricane Sandy-related service provision. Also, as in prior studies, mental health symptoms were evaluated using self-reported measures that are not diagnostic in nature, but result in assessments of symptomatology. These measures, however, have repeatedly demonstrated validity and reliability and are consistently correlated with more comprehensive diagnostic tools assessing their corresponding mental health disorder and mental health symptoms [25,26,48–50]. Another limitation was the low response rate to the second wave of surveys. While efforts were made to increase the ease and accessibility of surveys for participants by providing the survey via email, the response rate was detrimentally impacted by the reality that participants declined to answer their phones. Additionally, participants that completed both waves of the survey were composed of more females and had higher levels of hurricane exposure compared to the remaining cohort that did not complete follow-up questionnaires. Despite the limitations, this study is the only one to date to use a population-based sample of people residing in Long Island, Staten Island and Queens with longitudinal follow-up after Hurricane Sandy. The contribution of longitudinal results may inform future long-term rescue and recovery efforts in the event of another natural disaster.

5. Conclusions

In summary, there were significant decreases in anxiety and PTSD, but not depression scores, between baseline and follow-up. The strongest predictor of experiencing any mental health symptom after a disaster, however, was having anxiety, depression, or PTSD symptoms at baseline. Property, personal, and overall hurricane exposures were associated with an increased risk of having PTSD symptoms at follow-up, but significant effects were not observed between any type of hurricane exposure and anxiety or depression symptoms. To provide appropriate assistance and treatment after future natural disasters, mental health care providers and policymakers must consider the long-term effects of hurricane exposure on mental health, especially on PTSD, and must also focus on the vulnerable subgroup of those with existing mental health concerns.

Supplementary Materials: The following are available online at www.mdpi.com/1660-4601/14/9/957/s1, Table S1: Description of the exposure measurements collected, Table S2: Spearman correlation matrix of continuous study variables.

Acknowledgments: Research reported in this presentation was supported by The Centers for Disease Control and Prevention under award number U01-TP000573-01; the Office of the Assistant Secretary for Preparedness and Response under HITEP150029-01-00; and The Association of State and Territorial Health Officials under ASTHO 64-45471.

Author Contributions: Emanuela Taioli and Rebecca M Schwartz conceptualized the study design and statistical analyses. Christina N Gillezeau collected data and conducted the statistical analysis. Bian Liu supervised the statistical analysis. Wil Lieberman-Cribbin conducted statistical analyses and manuscript writing.

Conflicts of Interest: The authors declare no conflicts of interest. The founding sponsors had no role in the design of the study; in the collection, analyses, or interpretation of data; in the writing of the manuscript; and in the decision to publish the results.

References

1. Hurricane Research Division. The Thirty Costliest Mainland Us Tropical Cyclones 1900–2013. National Oceanic & Atmostpheric Administration, 2014. Available online: http://www.aoml.noaa.gov/hrd/tcfaq/costliesttable.html (accessed on 6 April 2016).
2. Centers for Disease Control and Prevention. Deaths associated with hurricane sandy—October–november 2012. *Morb. Mortal. Wkly. Rep.* **2013**, *62*, 393–397.
3. Staff, C.W. Superstorm Sandy: By the Numbers. Available online: http://www.cnn.com/2012/10/30/us/sandy-by-the-numbers/ (accessed on 3 April 2016).

4. Abramson, D.; Alst, D.V.; Beedasy, J.; Findley, P.; Peek, L. Person report: The Sandy Child & Family Health Study. Columbia University National Center for Disaster Preparedness; New York University College of Global Public Health; Rutgers School of Social Work, 2015; Available online: https://www.researchgate.net/publication/280925388_Sandy_Child_and_Family_Health_Study_PERSON_Report (accessed on 6 April 2016).
5. Boscarino, J.A. Community disasters, psychological trauma, and crisis intervention. *Int. J. Emerg. Mental Health* **2015**, *17*, 369–371.
6. Caramanica, K.; Brackbill, R.M.; Stellman, S.D.; Farfel, M.R. Posttraumatic stress disorder after hurricane sandy among persons exposed to the 9/11 disaster. *Int. J. Emerg. Mental Health* **2015**, *17*, 356–362. [CrossRef]
7. Galea, S.; Nandi, A.; Vlahov, D. The epidemiology of post-traumatic stress disorder after disasters. *Epidemiol. Rev.* **2005**, *27*, 78–91. [CrossRef] [PubMed]
8. Neria, Y.; Shultz, J.M. Mental health effects of hurricane sandy: Characteristics, potential aftermath, and response. *JAMA* **2012**, *308*, 2571–2572. [CrossRef] [PubMed]
9. Rhodes, J.; Chan, C.; Paxson, C.; Rouse, C.E.; Waters, M.; Fussell, E. The impact of hurricane katrina on the mental and physical health of low-income parents in new orleans. *Am. J. Orthopsychiatry* **2010**, *80*, 237. [CrossRef] [PubMed]
10. Galea, S.; Tracy, M.; Norris, F.; Coffey, S.F. Financial and social circumstances and the incidence and course of ptsd in mississippi during the first two years after hurricane katrina. *J. Trauma. Stress* **2008**, *21*, 357–368. [CrossRef] [PubMed]
11. Kessler, R.C.; Galea, S.; Gruber, M.J.; Sampson, N.A.; Ursano, R.J.; Wessely, S. Trends in mental illness and suicidality after hurricane katrina. *Mol. Psychiatry* **2008**, *13*, 374–384. [CrossRef] [PubMed]
12. Paxson, C.; Fussell, E.; Rhodes, J.; Waters, M. Five years later: Recovery from post traumatic stress and psychological distress among low-income mothers affected by hurricane katrina. *Soc. Sci. Med.* **2012**, *74*, 150–157. [CrossRef] [PubMed]
13. Heid, A.R.; Christman, Z.; Pruchno, R.; Cartwright, F.P.; Wilson-Genderson, M. Vulnerable, but why? Post-traumatic stress symptoms in older adults exposed to hurricane sandy. *Disaster Med. Public Health Prep.* **2016**, *10*, 362–370. [CrossRef] [PubMed]
14. Heid, A.R.; Pruchno, R.; Cartwright, F.P.; Wilson-Genderson, M. Exposure to hurricane sandy, neighborhood collective efficacy, and post-traumatic stress symptoms in older adults. *Aging Ment. Health* **2016**, *21*, 742–750. [CrossRef] [PubMed]
15. Hamama-Raz, Y.; Palgi, Y.; Shrira, A.; Goodwin, R.; Kaniasty, K.; Ben-Ezra, M. Gender differences in psychological reactions to hurricane sandy among new york metropolitan area residents. *Psychiat. Quart.* **2015**, *86*, 285–296. [CrossRef] [PubMed]
16. Palgi, Y.; Shrira, A.; Hamama-Raz, Y.; Palgi, S.; Goodwin, R.; Ben-Ezra, M. Not so close but still extremely loud: Recollection of the world trade center terror attack and previous hurricanes moderates the association between exposure to hurricane sandy and posttraumatic stress symptoms. *Compr. Psychiatry* **2014**, *55*, 807–812. [CrossRef] [PubMed]
17. Lowe, S.R.; Sampson, L.; Gruebner, O.; Galea, S. Mental health service need and use in the aftermath of hurricane sandy: Findings in a population-based sample of new york city residents. *Community Ment. Health J.* **2015**, *52*, 25–31. [CrossRef] [PubMed]
18. Manuel, J. The long road to recovery: Environmental health impacts of hurricane sandy. *Environ. Health Perspect.* **2013**, *121*, a152. [CrossRef] [PubMed]
19. Gruebner, O.; Lowe, S.R.; Sampson, L.; Galea, S. The geography of post-disaster mental health: Spatial patterning of psychological vulnerability and resilience factors in new york city after hurricane sandy. *Int. J. Health Geogr.* **2015**, *14*, 16. [CrossRef] [PubMed]
20. Schwartz, R.M.; Rothenberg, P.; Kerath, S.M.; Liu, B.; Taioli, E. The lasting mental health effects of hurricane sandy on residents of the rockaways. *J. Emerg. Manag.* **2016**, *14*, 269–279. [CrossRef] [PubMed]
21. Kim, H.; Schwartz, R.M.; Hirsch, J.; Silverman, R.; Liu, B.; Taioli, E. Effect of hurricane sandy on long island emergency departments visits. *Disaster Med. Public Health Prep.* **2016**, *10*, 344–350. [CrossRef] [PubMed]
22. Schwartz, R.; Liu, B.; Sison, C.; Kerath, S.M.; Breil, T.; Murphy, L.; Taioli, E. Study design and results of a population-based study on perceived stress following hurricane sandy. *Disaster Med. Public Health Prep.* **2016**, *10*, 325–332. [CrossRef] [PubMed]

23. Schwartz, R.M.; Sison, C.; Kerath, S.M.; Murphy, L.; Breil, T.; Sikavi, D.; Taioli, E. The impact of hurricane sandy on the mental health of new york area residents. *Am. J. Disaster Med.* **2014**, *10*, 339–346. [CrossRef] [PubMed]

24. Lieberman-Cribbin, W.; Liu, B.; Schneider, S.; Schwartz, R.; Taioli, E. Self-reported and fema flood exposure assessment after hurricane sandy: Association with mental health outcomes. *PLoS ONE* **2017**, *12*, e0170965. [CrossRef] [PubMed]

25. Kroenke, K.; Spitzer, R.L.; Williams, J.B.; Lowe, B. An ultra-brief screening scale for anxiety and depression: The phq-4. *Psychosomatics* **2009**, *50*, 613–621. [PubMed]

26. Ruggiero, K.J.; Del Ben, K.; Scotti, J.R.; Rabalais, A.E. Psychometric properties of the ptsd checklist-civilian version. *J. Trauma. Stress* **2003**, *16*, 495–502. [CrossRef] [PubMed]

27. Health, U.D.O.; Services, H. *US Department of Agriculture. Dietary Guidelines for Americans 2015–2020*; Skyhorse Publishing Inc.: New York, NY, USA, 2016.

28. Galea, S.; Brewin, C.R.; Gruber, M.; Jones, R.T.; King, D.W.; King, L.A.; McNally, R.J.; Ursano, R.J.; Petukhova, M.; Kessler, R.C. Exposure to hurricane-related stressors and mental illness after hurricane katrina. *Arch Gen. Psychiatry* **2007**, *64*, 1427–1434. [CrossRef] [PubMed]

29. Lowe, S.R.; Sampson, L.; Gruebner, O.; Galea, S. Community unemployment and disaster-related stressors shape risk for posttraumatic stress in the longer-term aftermath of hurricane sandy. *J Trauma. Stress* **2016**, *29*, 440–447. [CrossRef] [PubMed]

30. Gill, D.A. Secondary trauma or secondary disaster? Insights from hurricane katrina. *Sociol. Spectrum* **2007**, *27*, 613–632. [CrossRef]

31. Chan, C.S.; Lowe, S.R.; Weber, E.; Rhodes, J.E. The contribution of pre-and postdisaster social support to short-and long-term mental health after hurricanes katrina: A longitudinal study of low-income survivors. *Soc. Sci. Med.* **2015**, *138*, 38–43. [CrossRef] [PubMed]

32. Lowe, S.R.; Rhodes, J.E. Trajectories of psychological distress among low-income, female survivors of hurricane katrina. *Am. J Orthopsychiat.* **2013**, *83*, 398–412. [CrossRef] [PubMed]

33. Glass, K.; Flory, K.; Hankin, B.L.; Kloos, B.; Turecki, G. Are coping strategies, social support, and hope associated with psychological distress among hurricane katrina survivors? *J. Soc. Clin. Psychol.* **2009**, *28*, 779–795. [CrossRef]

34. Pina, A.A.; Villalta, I.K.; Ortiz, C.D.; Gottschall, A.C.; Costa, N.M.; Weems, C.F. Social support, discrimination, and coping as predictors of posttraumatic stress reactions in youth survivors of hurricane katrina. *J. Clin. Child Adolesc. Psychol.* **2008**, *37*, 564–574. [CrossRef] [PubMed]

35. Hobfoll, S.E. Conservation of resources: A new attempt at conceptualizing stress. *Am. Psychol.* **1989**, *44*, 513–524. [CrossRef] [PubMed]

36. Hobfoll, S.E. Social and psychological resources and adaptation. *Rev. Gen. Psychol.* **2002**, *6*, 307. [CrossRef]

37. Benight, C.C.; Ironson, G.; Klebe, K.; Carver, C.S.; Wynings, C.; Burnett, K.; Greenwood, D.; Baum, A.; Schneiderman, N. Conservation of resources and coping self-efficacy predicting distress following a natural disaster: A causal model analysis where the environment meets the mind. *Anxiety Stress Coping* **1999**, *12*, 107–126. [CrossRef]

38. Bonanno, G.A.; Galea, S.; Bucciarelli, A.; Vlahov, D. What predicts psychological resilience after disaster? The role of demographics, resources, and life stress. *J. Consult. Clin. Psychol.* **2007**, *75*, 671. [CrossRef] [PubMed]

39. Freedy, J.R.; Saladin, M.E.; Kilpatrick, D.G.; Resnick, H.S.; Saunders, B.E. Understanding acute psychological distress following natural disaster. *J. Trauma. Stress* **1994**, *7*, 257–273. [CrossRef] [PubMed]

40. Norris, F.H.; Perilla, J.L.; Riad, J.K.; Kaniasty, K.; Lavizzo, E.A. Stability and change in stress, resources, and psychological distress following natural disaster: Findings from hurricane andrew. *Anxiety Stress Coping* **1999**, *12*, 363–396. [CrossRef] [PubMed]

41. Hobfoll, S.E. Conservation of resources and disaster in cultural context: The caravans and passageways for resources. *Psychiatry* **2012**, *75*, 227–232. [CrossRef] [PubMed]

42. Wadsworth, M.E.; Santiago, C.D.; Einhorn, L. Coping with displacement from hurricane katrina: Predictors of one-year post-traumatic stress and depression symptom trajectories. *Anxiety Stress Coping* **2009**, *22*, 413–432. [CrossRef] [PubMed]

43. Zwiebach, L.; Rhodes, J.; Roemer, L. Resource loss, resource gain, and mental health among survivors of hurricane katrina. *J. Trauma. Stress* **2010**, *23*, 751–758. [CrossRef] [PubMed]

44. Ehrlich, M.; Harville, E.; Xiong, X.; Buekens, P.; Pridjian, G.; Elkind-Hirsch, K. Loss of resources and hurricane experience as predictors of postpartum depression among women in Southern Louisiana. *J. Womens Health* **2010**, *19*, 877–884. [CrossRef] [PubMed]

45. Sampson, L.; Lowe, S.R.; Gruebner, O.; Cohen, G.H.; Galea, S. Perceived service need after hurricane sandy in a representative sample of survivors: The roles of community-level damage and individual-level stressors. *Disaster Med. Public Health Prep.* **2016**, *10*, 428–435. [CrossRef] [PubMed]

46. Sullivan, G.; Vasterling, J.J.; Han, X.; Tharp, A.T.; Davis, T.; Deitch, E.A.; Constans, J.I. Preexisting mental illness and risk for developing a new disorder after hurricane katrina. *J. Nerv. Mental Dis.* **2013**, *201*, 161–166. [CrossRef] [PubMed]

47. Lowe, S.R.; Sampson, L.; Gruebner, O.; Galea, S. Psychological resilience after hurricane sandy: The influence of individual-and community-level factors on mental health after a large-scale natural disaster. *PLoS ONE* **2015**, *10*, e0125761. [CrossRef] [PubMed]

48. Weathers, F.; Litz, B.; Herman, D.; Huska, J.A.; Keane, T.M. The PTSD checklist (PCL): Reliability, validity, and diagnostic utility. In Proceedings of the Annual Convention of the International Society for Traumatic Stress Studies, San Antonio, TX, USA, 25 October 1993.

49. Blanchard, E.B.; Jones-Alexander, J.; Buckley, T.C.; Forneris, C.A. Psychometric properties of the PTSD checklist (PCL). *Behav. Res. Ther.* **1996**, *34*, 669–673. [CrossRef]

50. Löwe, B.; Wahl, I.; Rose, M.; Spitzer, C.; Glaesmer, H.; Wingenfeld, K.; Schneider, A.; Brähler, E. A 4-item measure of depression and anxiety: Validation and standardization of the patient health questionnaire-4 (PHQ-4) in the general population. *J. Affect. Disorders* **2010**, *122*, 86–95. [CrossRef] [PubMed]

International Journal of
*Environmental Research
and Public Health*

MDPI

Communication

Preliminary Assessment of Hurricane Harvey Exposures and Mental Health Impact

Rebecca M. Schwartz [1,2,3], Stephanie Tuminello [2,3], Samantha M. Kerath [1,4], Janelle Rios [5], Wil Lieberman-Cribbin [2,3] and Emanuela Taioli [2,3,*]

[1] Department of Occupational Medicine, Epidemiology and Prevention, Hofstra Northwell Health School of Medicine, Great Neck, NY 11021, USA; Rschwartz3@northwell.edu (R.M.S.); Skerath@northwell.edu (S.M.K.)

[2] Department of Population Health Science and Policy and Institute for Translational Epidemiology, Icahn School of Medicine at Mount Sinai, New York, NY 10029, USA; Stephanie.Tuminello@mssm.edu (S.T.); wil.lieberman-cribbin@icahn.mssm.edu (W.L.C.)

[3] Center for Disaster Health, Trauma and Resilience; Mount Sinai, Stony Brook University, Northwell Health, Stony Brook, NY 11794, USA

[4] Center for Biomedical Science, Feinstein Institute for Medical Research, Manhasset, NY 11030, USA

[5] The University of Texas School of Public Health, Houston, TX 75235, USA; Janelle.Rios@uth.tmc.edu

[*] Correspondence: Emanuela.Taioli@mountsinai.org

Received: 10 April 2018; Accepted: 9 May 2018; Published: 13 May 2018

Abstract: Hurricane Harvey made landfall in Houston, Texas on 25 August 2017, the psychological and physical effects of which are still unknown. We assessed hurricane exposure and the immediate mental health needs of the population to define public health priorities for a larger epidemiological study. Convenience sampling was used to recruit participants ($n = 41$) from the greater Houston area aged ≥ 18 years. Participants completed a questionnaire about demographics, hurricane exposures, and physical/mental health. Post-Traumatic Stress Disorder (PTSD) was measured with the Post-Traumatic Stress Disorder Checklist-S (PCL-S; a score ≥ 30 indicated probable PTSD symptoms). The Patient Health Questionnaire-4 (PHQ-4) was used to assess symptoms of depression and generalized anxiety disorder. The average PTSD score was 32.9 (SD = 17.1); a total of 46% of participants met the threshold for probable PTSD. Increased overall hurricane exposure (adjusted odds ratio (OR_{adj}) 1.42; 95% confidence interval (CI): 1.06–2.05) and property-related exposure (OR_{adj} 1.53; 95% CI: 1.07–2.18) were both statistically significantly associated with increased odds of probable PTSD symptoms. A perception of chemical/toxin exposure due to Hurricane Harvey was reported by 44% of participants. A higher number of personal or property exposures were associated with greater mental health symptoms three weeks post-hurricane. This work has implications for the ongoing response to Hurricane Harvey and for assessing the immediate needs of the population.

Keywords: extreme weather event; disaster; post-traumatic stress disorder; emergency response; epidemiology

1. Introduction

At 10 PM on 25 August 2017, Hurricane Harvey hit the coast of Texas as a Category 4 hurricane with wind speeds greater than 58 m/s. Between landfalls over a four day period, Hurricane Harvey brought torrential rains totaling about 50 inches at Bush International Airport in Houston, Texas [1]. This storm was particularly devastating to the greater Houston area, a city with 2.3 million people. Various sources and media outlets indicated 30,000 residents displaced, $70–$170 billion in property damage, and half-million vehicles and untold structures flooded.

Hurricane exposure has been shown to have a profound impact on the mental health of affected residents, leading to increased symptomology of stress, anxiety, depression and Post-Traumatic Stress Disorder (PTSD) [2–4]. Being displaced because of the storm, especially to a temporary shelter, has been shown to exacerbate these mental health issues, possibly because of the perceived decrease in social support associated with displacement [5–9].

During Hurricane Harvey there was also the potential for chemical exposures due to numerous chemical plants, natural gas and oil refineries present in the Houston area. More than 50 Environmental Protection Agency (EPA) Superfund sites (sites of previous contamination) are located in the area, some of which were flooded and potentially contaminated the flood waters in their surroundings [10]. Further, anecdotal exposure to mold and sewage bacteria in flooded streets was widely reported. Exposures related specifically to chemicals can have both short and long-term negative impacts on mental health [11–14]. Previous research indicates an indirect impact of chemical disasters on mental health. After the Deepwater Horizon Oil Spill, participants who experienced job loss and disruption in social and occupational functioning as a result of the oil spill were more likely to experience anxiety and depression [11,12]. A study of South Carolina residents following a chlorine gas disaster indicated that physical health symptoms were significantly associated with post-traumatic stress, suggestive of an interaction between physical and mental health during exposures to chemicals and toxins [13]. Similarly, dampness and mold were associated with depression, but the association was mediated by perception of control over one's home and by physical health [14]. Another issue related to chemical exposures is cognitive bias; the perceived odor and cognitive expectations of a chemical can negatively affect how an individual responds to that chemical [15]. The impact of chemical exposures on the physical and mental health of an affected population becomes even more pronounced when combined with other psychological stressors related to hurricanes, such as escaping flooded homes, finding shelter and temporary housing, or replacing damaged property [16]. Anecdotal reports from healthcare providers hint at these psychological and physical effects; however, the full effects of both the actual chemical contamination and the indirect effects of perceived exposures due to Hurricane Harvey are still being elucidated.

We conducted a preliminary assessment of the initial psychological impact of Hurricane Harvey in an effort to assess the immediate mental health needs of the population and to define public health priorities for a larger epidemiological study. Proximity in time to Hurricane Harvey was an important consideration in our research planning as the immediate weeks following a natural disaster are typically when the mental health impacts are arguably the most severe [17,18]. To our knowledge we were one of the first research teams to reach the Houston area following Hurricane Harvey.

2. Materials and Methods

Our research team was in Houston less than 3-weeks after Hurricane Harvey made landfall, which was the quickest our team could mobilize, and immediately began surveying affected residents. Convenience sampling was used to recruit participants from heavily affected areas including the George R. Brown convention center which, in the immediate aftermath of Hurricane Harvey, was converted into a temporary shelter. Participants who were ≥18 years old and had resided within the greater Houston area during the hurricane were eligible and were given a $10 Target gift card as reimbursement for their time. Most found out about the study through word of mouth. Approval for this study was given by the internal review board of the Icahn School of Medicine at Mount Sinai on 9/11/2017 (HS#: 15-00513) as a modification of our existing IRB created for researching the effects of Hurricane Sandy.

2.1. Hurricane Exposures

Study participants were consented and completed a questionnaire about demographics, hurricane exposures, and physical/mental health before and after Hurricane Harvey, modelled on studies we conducted on Sandy affected populations [2,19]. Personal hurricane exposures were those that directly affected the participant or their family, and property related exposures were exposures related to the level of personal property affected and the resulting financial hardship (Supplementary Table S1) [19]. The grouping of an exposure item into either the Personal or Property-related category was based on the results of a Principal Components Analysis conducted by Schwartz et al. [19]. There was a total of 16 personal exposures and 14 property exposures measured. A "total exposure score" was also generated by summing affirmative answers (1 = Yes) to each of the 30 items (personal and property) on the hurricane exposure scale. Displacement, one of the personal exposure items, was also examined separately to determine the mental health correlates of displacement specifically.

Participants also reported exposure to specific chemicals and toxins (Yes/No) known to be associated with Harvey, such as debris, mold, petroleum, and chemical emissions (e.g., carbon monoxide).

2.2. Mental Health Outcomes

The primary outcomes were mental health symptoms of anxiety, depression and PTSD. PTSD symptoms were examined using the Post-Traumatic Stress Disorder Checklist-S (PCL-S), a 17-item self-report measure that asked about PTSD symptoms specific to Hurricane Harvey. A score ≥ 30 was considered indicative of probable PTSD symptoms. The Patient Health Questionnaire-4 (PHQ-4) was used to assess symptoms of depression and of generalized anxiety disorder; a score ≥ 3 was considered indicative of probable depression or anxiety.

2.3. Statistical Analysis

Fisher's Exact Test or Wilcoxon Rank Sum and multivariate logistic regression were used to evaluate associations between hurricane exposures and mental health. Having probable PTSD, anxiety or depression was the primary outcome in the logistic regression statistical models, and was treated as a dichotomous variable (yes or no) based on the clinically relevant cutoffs described above. The number of personal, property-related or chemical exposures was used as a continuous variable to investigate how an increase in any type of hurricane exposure affected mental health. Logistic models were adjusted for covariates including age, gender and a prior history of a mental health condition, including anxiety disorder, depression, PTSD, schizophrenia, bipolar disorder, substance abuse disorders (alcohol or prescription drug related) or some other mental health disorder that was diagnosed by a physician. We chose to include these covariates because of their perceived biological significance, which is supported by the literature [20]. Health insurance status was also included in the adjusted models as it was found to be statistically associated with both PTSD and depression in the bivariate analysis (data not shown), and because it acts as a proxy for socioeconomic status. Though other variables such as race, ethnicity and education might still have a confounding effect, our data represents only a small preliminary assessment of the affected Houston population and we did not have sufficient sample size to run logistic models with adjustment for all these covariates. Data analysis was performed using SAS (SAS Institute, Cary, NC, USA, V9.4) software.

3. Results

The study sample consisted of 41 participants who had lived in the greater Houston area during Hurricane Harvey. The majority was female (56%) US born (78%), and had some type of health insurance (78%). The study cohort was mostly White (34%), Black (32%) and Hispanic (20%); the remaining participants self-identified as American Indian, Pacific Islander, Asian or Other. The majority of participants had attended at least some college (76%). The mean age was 44 years (SD = 10 years), with a mean household size of 2.4 people (Table 1).

Table 1. Characteristics of the population under study.

Demographics	N = 41
Gender	
Female	23 (56.1%)
Male	18 (41.9%)
Ethnicity	
White	14 (35.0%)
Black	13 (32.5%)
American Indian	1 (2.5%)
Hispanic	8 (20.0%)
Pacific Islander/Asian	4 (10.0%)
US Born	
No	9 (22.0%)
Yes	32 (78.0%)
Education	
<High School Degree	3 (7.9%)
High School Degree	6 (15.8%)
Some College	8 (21.0%)
College Degree	8 (21.1%)
Post-Graduate Degree	13 (34.2%)
Health Insurance	
No	9 (22.0%)
Yes	32 (78.0%)
Mental Health Condition Prior to Harvey	
No	29 (70.7%)
Yes	12 (29.3%)
Age (years)	44.2 ± 10.3
Household Size (# of people)	2.4 ± 1.1

Frequency missing: Ethnicity ($n = 1$), Education ($n = 3$), Age ($n = 3$), Household Size ($n = 1$).

3.1. Hurricane Exposures

Most participants (88%) reported experiencing some type of exposure to Hurricane Harvey. Overall, 34% of study participants reported at least one type of Personal exposure, while 61% reported at least one type of Property-related exposure. The most commonly reported personal exposure items were assisting in rescue efforts (31.7%) and being evacuated from their homes (22.0%). For Property-related exposure, participants most commonly reported being displaced (53.7%), having their homes damaged (53.7%), and having flooding in their homes (46.3%; Figure 1).

A perception of chemical/toxin exposure due to Hurricane Harvey was also reported by a large number of participants (44%): 39% reported being exposed to dirty or contaminated flood water, 27% to sewage, and 12% to oil leaks. Participants in this study also reported being exposed to mold (24.4%) and debris (31.7%; Figure 1).

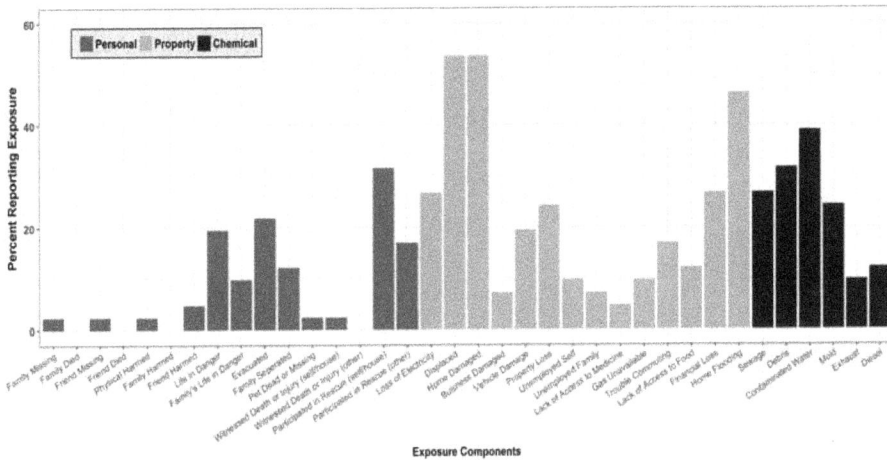

Figure 1. Hurricane Exposure Assessment (n = 41). The mean number of overall hurricane exposures was 4.19 (SD = 4.56; range 0–30).

3.2. Mental Health Symptomology

PCL scores ranged from 17–71, with a mean score of 32.9 (SD = 17.1); 46% of participants met the threshold for probable PTSD symptoms. As determined by the PHQ-4, 53.7% of participants experienced anxiety symptoms and 39.0% experienced depression symptoms post-Harvey.

After adjustment, increased overall hurricane exposure (adjusted odds ratio (OR_{adj}) 1.42; 95% confidence interval (CI): 1.06–2.05) and property-related exposure (OR_{adj} 1.53; 95% CI: 1.07–2.18) were both statistically significantly associated with an increased odds of probable PTSD symptoms. Further, being displaced during the hurricane was also significantly associated with increased odds of probable PTSD symptoms (OR_{adj} 12.50; 95% CI: 1.43–108.92).

After adjustment, an increase in chemical/toxin exposure (OR_{adj} 1.96; 95% CI: 1.16–3.32), overall hurricane exposure (OR_{adj} 1.50; 95% CI: 1.09–2.06), and property-related hurricane exposure (OR_{adj} 1.60; 95% CI: 1.13–2.28) were all statistically significantly associated with increased odds of probable anxiety.

In terms of probable depression symptoms, only the relationship with displacement remained significant after adjustment (OR_{adj} 15.76; 95% CI: 1.22–203.06) (Table 2).

IJERPH 2018, 15, 974

Table 2. Association between Hurricane Harvey exposure and mental health symptoms.

Exposure	PTSD (PCL ≥ 30) Mean (SD)			Anxiety (PHQ-4 ≥ 3) Mean (SD)			Depression (PHQ-4 ≥ 3) Mean (SD)		
	No	Yes	OR_{adj} * (95% CI)	No	Yes	OR_{adj} * (95% CI)	No	Yes	OR_{adj} * (95% CI)
Chemical/Toxin	0.77 (1.45)	2.21 (2.35) ^	1.45 (0.95–2.21) ^	0.53 (1.12)	2.23 (2.31) +	1.96 (1.16–3.32)	1.04 (1.85)	2.06 (2.21)	1.36 (0.94–1.96)
Hurricane overall	2.36 (2.32)	6.95 (5.29) ^	1.42 (1.06–2.05)	2.36 (2.34)	6.33 (5.22) ^	1.50 (1.09–2.06)	3.72 (4.66)	5.69 (4.25) @	1.10 (0.93–1.31)
Hurricane Personal	0.68 (0.84)	2.00 (2.26) #	2.29 (0.92–5.71)	0.79 (0.92)	1.73 (2.19)	1.97 (0.92–4.27)	1.20 (1.68)	1.44 (1.93)	1.10 (0.72–1.68)
Hurricane Property	1.68 (1.94)	4.95 (3.44)	1.53 (1.07–2.18)	1.58 (1.92)	4.59 (3.39) ^	1.60 (1.13–2.28)	2.52 (3.28)	4.25 (2.74) +	1.18 (0.93–1.51)
Displacement (%)									
No	17 (89)	2 (11) ^	1 (ref)	13 (68)	6 (32) ^	1 (ref)	16 (84)	3 (16) ^	1 (ref)
Yes	5 (23)	17 (77)	12.50 (1.43–108.92)	6 (27)	16 (73)	6.66 (0.922–48.09)	9 (41)	13 (59)	15.76 (1.22–203.06)

* adjusted for age (continuous), gender, health insurance status and previous history of mental health conditions; n = 39, 2 people were missing age; ^ $p < 0.01$; # $p = 0.04$; + $p = 0.02$; @ $p = 0.03$.

48

4. Discussion

4.1. Hurricane Exposure and Mental Health

Preliminary results indicate that increased Hurricane Harvey exposure may have had a significant impact on the mental health of Houston residents. The reported number of personal or property exposures was positively associated with mental health symptoms, particularly PTSD symptoms, three weeks post-hurricane. Further, consistent with our previous work on displacement during Hurricane Sandy, the current study indicated that displacement was associated with an increased risk of PTSD and depression symptoms [19,21].

The data also shows an association between perceived chemical/toxin hurricane exposure and PTSD and anxiety symptoms. The results support previous research reporting that exposures related specifically to chemicals have negative short and long-term impacts on mental health [11,12,22]. However, the interaction between physical and mental health following exposure to toxins/chemicals after a natural disaster is still largely unknown and should be the focus of further studies, including a comparison of objective versus perceived measurement of chemical exposure.

4.2. Strengths and Limitations

This study was based on a small convenience sample; it is cross-sectional, and data interpretation was correlational with no causal inferences made. However, to our knowledge, this represents the first reported assessment of Hurricane Harvey's impact on Houston residents' mental health. It incorporated validated instruments to investigate the degree of hurricane exposure and its impact on mental health symptoms. It should be noted that it is not possible to receive a true diagnosis of PTSD only three weeks after a traumatic event, however the PCL-S is used to assess symptoms and is not used to make a true diagnosis. Further, it is possible that the PTSD scale is assessing acute stress symptoms, as opposed to potential PTSD symptoms, given the close proximity to the hurricane. Recall bias, a potential limitation of this study design, was minimized by the quick mobilization of the research team.

This study's proximity in time to the hurricane caused limitations to our recruitment. Being in Houston so soon after the hurricane meant that we faced physical barriers such as flooded and otherwise inaccessible roads. However, this also enabled us to recruit participants while they were still displaced to a shelter. The need to seek a temporary shelter most likely compounds the hurricane-related mental health symptoms caused or exacerbated by the storm [2,6,23]. Moreover, having a pre-existing mental health condition has been linked to greater mental health symptomology post-hurricane, and those seeking temporary shelter are often unable to retrieve their necessary psychiatric medication during an evacuation [18,24]. The fact that we do not see a similar statistically significant relationship in our study between pre-existing mental health disorders and mental health symptoms post-hurricane can most likely be attributed to the study's small sample size. Research directly after the hurricane is necessary to assess the mental health needs of displaced residents.

4.3. Future Research Implications

The lessons gained from conducting this preliminary assessment of Houston residents immediately after Hurricane Harvey has been instrumental in our future research endeavors. One of the most significant research barriers we faced during this initial assessment was receiving permission to be at the shelter as a research group. We were unsure who the designated decision-maker at the shelter was and what protocol we needed to follow to allow us access to the shelter for research purposes. Though we were able to overcome these obstacles, it speaks to the great need to have community partnerships and infrastructure in place prior to disasters [25]. After Hurricane Sandy, community engagement with those in the Rockaway area of New York allowed us to link those affected by the hurricane to local mental health resources including an existing mental health and substance abuse treatment center in the area. Hopefully, with guidance from this assessment and future research, similar programs will be able to be organized in Houston.

5. Conclusions

This assessment of the hurricane's initial impact on mental health lays the groundwork for future research. It is clear that residents were greatly impacted by the hurricane and that the types of exposures they suffered were vast and variable. These findings have implications both for the ongoing response to Hurricane Harvey and the emergency preparedness community as a whole. Literature points to the stability of PTSD over time among communities exposed to natural disasters, with rates as high as 30–40% [4], indicating the need to address symptoms as early as possible and to provide long-term support for those affected.

Supplementary Materials: The following are available online at http://www.mdpi.com/1660-4601/15//974/s1, Table S1: Personal vs. Property Hurricane Exposure Items.

Author Contributions: E.T. and R.M.S. conceptualized the study design and statistical analyses. S.T., S.M.K. and J.R. collected data. S.T. and W.L.-C. conducted statistical analyses and manuscript writing.

Acknowledgments: This work was partly supported by departmental funding and the Northwell Health and Stony Brook—Mount Sinai Alliance Pilot Project.

Conflicts of Interest: The authors declare no conflict of interest.

References

1. CNN, B.G. CNN Design: Lansing Cai Harvey's Devastating Impact by the Numbers. Available online: http://www.cnn.com/2017/08/27/us/harvey-impact-by-the-numbers-trnd/index.html (accessed on 25 October 2017).
2. Schwartz, R.M.; Sison, C.; Kerath, S.M.; Murphy, L.; Breil, T.; Sikavi, D.; Taioli, E. The impact of Hurricane Sandy on the mental health of New York area residents. *Am. J. Disaster Med.* **2015**, *10*, 339–346. [CrossRef] [PubMed]
3. Cerdá, M.; Bordelois, P.M.; Galea, S.; Norris, F.; Tracy, M.; Koenen, K.C. The course of posttraumatic stress symptoms and functional impairment following a disaster: What is the lasting influence of acute versus ongoing traumatic events and stressors? *Soc. Psychiatry Psychiatr. Epidemiol.* **2013**, *48*, 385–395. [CrossRef] [PubMed]
4. Neria, Y.; Nandi, A.; Galea, S. Post-traumatic stress disorder following disasters: A systematic review. *Psychol. Med.* **2008**, *38*, 467–480. [CrossRef] [PubMed]
5. Schwartz, R.M.; Liu, B.; Lieberman-Cribbin, W.; Taioli, E. Displacement and mental health after natural disasters. *Lancet Planet. Health* **2017**, *1*, e314. [CrossRef]
6. Lê, F.; Tracy, M.; Norris, F.H.; Galea, S. Displacement, county social cohesion, and depression after a large-scale traumatic event. *Soc Psychiatry Psychiatr. Epidemiol.* **2013**, *48*, 1729–1741. [CrossRef] [PubMed]
7. Schwartz, R.M.; Rasul, R.; Kerath, S.M.; Watson, A.R.; Lieberman-Cribbin, W.; Liu, B.; Taioli, E. Displacement during Hurricane Sandy: The impact on mental health. *J. Emerg. Manag.* **2018**, *16*, 17–27. [CrossRef] [PubMed]
8. Fullilove, M.T. Psychiatric implications of displacement: Contributions from the psychology of place. *Am. J. Psychiatry* **1996**, *153*, 1516–1523. [CrossRef] [PubMed]
9. Wadsworth, M.E.; Santiago, C.D.; Einhorn, L. Coping with displacement from Hurricane Katrina: Predictors of one-year post-traumatic stress and depression symptom trajectories. *Anxiety Stress Coping* **2009**, *22*, 413–432. [CrossRef] [PubMed]
10. CNBC Hurricane Harvey Rains Flood Toxic Superfund Sites in Texas. Available online: https://www.cnbc.com/2017/09/03/hurricane-harvey-rains-flood-toxic-superfund-sites-in-texas.html (accessed on 25 October 2017).
11. Grattan, L.M.; Roberts, S.; Mahan, W.T.; McLaughlin, P.K.; Otwell, W.S.; Morris, J.G. The early psychological impacts of the Deepwater Horizon oil spill on Florida and Alabama communities. *Environ. Health Perspect.* **2011**, *119*, 838–843. [CrossRef] [PubMed]

12. Osofsky, H.J.; Osofsky, J.D.; Hansel, T.C. Deepwater horizon oil spill: Mental health effects on residents in heavily affected areas. *Disaster Med. Public Health Prep.* **2011**, *5*, 280–286. [CrossRef] [PubMed]

13. Clark, K.A.; Chanda, D.; Balte, P.; Karmaus, W.J.; Cai, B.; Vena, J.; Lawson, A.B.; Mohr, L.C.; Gibson, J.J.; Svendsen, E.R. Respiratory symptoms and lung function 8–10 months after community exposure to chlorine gas: A public health intervention and cross-sectional analysis. *BMC Public Health* **2013**, *13*, 945. [CrossRef] [PubMed]

14. Shenassa, E.D.; Daskalakis, C.; Liebhaber, A.; Braubach, M.; Brown, M. Dampness and Mold in the Home and Depression: An Examination of Mold-Related Illness and Perceived Control of One's Home as Possible Depression Pathways. *Am. J. Public Health* **2007**, *97*, 1893–1899. [CrossRef] [PubMed]

15. Dalton, P.; Wysocki, C.J.; Brody, M.J.; Lawley, H.J. The influence of cognitive bias on the perceived odor, irritation and health symptoms from chemical exposure. *Int. Arch. Occup. Environ. Health* **1997**, *69*, 407–417. [CrossRef] [PubMed]

16. Peek, M.; Cutchin, M.; Freeman, D.; Stowe, R.; Goodwin, J. Environmental hazards and stress: Evidence from the Texas City Stress and Health Study. *J. Epidemiol. Community Health* **2009**, *63*, 792–798. [CrossRef] [PubMed]

17. Flory, K.; Kloos, B.; Hankin, B.L.; Cheely, C.A. Clinical Research After Catastrophic Disasters: Lessons Learned From Hurricane Katrina. *Prof. Psychol. Res. Pract.* **2008**, *39*, 107–112. [CrossRef] [PubMed]

18. Ochi, S.; Hodgson, S.; Landeg, O.; Mayner, L.; Murray, V. Disaster-driven evacuation and medication loss: A systematic literature review. *PLoS Curr.* **2014**, *6*. [CrossRef] [PubMed]

19. Schwartz, R.; Liu, B.; Sison, C.; Kerath, S.M.; Breil, T.; Murphy, L.; Taioli, E. Study Design and Results of a Population-Based Study on Perceived Stress Following Hurricane Sandy. *Disaster Med. Public Health Prep.* **2016**, *10*, 325–332. [CrossRef] [PubMed]

20. Schwartz, R.M.; Gillezeau, C.N.; Liu, B.; Lieberman-Cribbin, W.; Taioli, E. Longitudinal Impact of Hurricane Sandy Exposure on Mental Health Symptoms. *Int. J. Environ. Res. Public Health* **2017**, *14*, 957. [CrossRef] [PubMed]

21. Rebecca, M.; Schwartz, P.; Patricia Rothenberg, B.A.; Samantha M. Kerath, M.S.; Bian Liu, P.; Emanuela Taioli, M.D. The lasting mental health effects of Hurricane Sandy on residents of the Rockaways. *J. Emerg. Manag.* **2016**, *14*, 269–279. [CrossRef]

22. Ginsberg, J.P.; Holbrook, J.R.; Chanda, D.; Bao, H.; Svendsen, E.R. Posttraumatic stress and tendency to panic in the aftermath of the chlorine gas disaster in Graniteville, South Carolina. *Soc. Psychiatry Psychiatr. Epidemiol.* **2012**, *47*, 1441–1448. [CrossRef] [PubMed]

23. Fussell, E.; Lowe, S.R. The impact of housing displacement on the mental health of low-income parents after Hurricane Katrina. *Soc. Sci. Med.* **2014**, *113*, 137–144. [CrossRef] [PubMed]

24. Sullivan, G.; Vasterling, J.J.; Han, X.; Tharp, A.T.; Davis, T.; Deitch, E.A.; Constans, J.I. Preexisting Mental Illness and Risk for Developing a New Disorder After Hurricane Katrina. *J. Nerv. Ment. Dis.* **2013**, *201*, 161–166. [CrossRef] [PubMed]

25. Springgate, B.F.; Wennerstrom, A.; Meyers, D.; Allen, C.E.; Vannoy, S.D.; Bentham, W.; Wells, K.B. Building community resilience through mental health infrastructure and training in post-Katrina New Orleans. *Ethn. Dis.* **2011**, *21*, S1–29. [PubMed]

International Journal of
*Environmental Research
and Public Health*

MDPI

Article

Weather and Suicide: A Decade Analysis in the Five Largest Capital Cities of Colombia

Julián Alfredo Fernández-Niño [1], Víctor Alfonso Flórez-García [1] ,

Claudia Iveth Astudillo-García [2,*] and Laura Andrea Rodríguez-Villamizar [3]

[1]   Departamento de Salud Pública, Universidad del Norte, Barranquilla (Atlántico) ZP 081007, Colombia;
     aninoj@uninorte.edu.co (J.A.F.-N.);vfloreza@uninorte.edu.co (V.A.F.-G.)
[2]   Servicios de Atención Psiquiátrica, Secretaría de Salud, Ciudad de México ZP 11410, Mexico
[3]   Departamento de Salud Pública, Universidad Industrial de Santander,
     Bucaramanga (Santander) ZP 68001, Colombia; laurovi@uis.edu.co
*   Correspondence: claudiaiveth.astudillo@gmail.com; Tel.: +52-5062-1600 (ext. 52335)

Received: 30 April 2018; Accepted: 6 June 2018; Published: 22 June 2018

Abstract: Historically, seasonal variations in suicide rates were thought to be associated with changes in weather. Most of this evidence however, is based on studies that were conducted in developed countries that are located outside the tropics. As such, it is necessary to examine this association in developing countries, such as Colombia, which do not experience marked seasons. In addition, it is important to adjust for the effect of holidays when analyzing this association as they have been reported to be a relevant confounding factor. Our objective was to estimate the association between daily suicide incidence among men and women in five major Colombian cities (Bogotá, Medellin, Cali, Barranquilla, and Bucaramanga) and daily temperature and rainfall. For this purpose, we conducted a multi-city, multi-temporal ecological study from 2005 to 2015, using data from the suicide mortality registries (provided by the National Administrative Department of Statistics). Daily measurements of the two weather variables were obtained from the official historical registry of the meteorological station at each city airport. We used these data to estimate conditional Poisson models for daily suicide counts, stratifying by sex and adjusting for holidays. Although we found that none of the weather variable estimators could reject the null hypothesis, we uncovered an association between suicide incidence and long weekends in the total suicide model (Incidence Rate Ratio (IRR): 1.19, 95% confidence interval (CI): 1.04–1.23). We found no evidence of association between weather variables and suicide in Colombia. Our study is based on daily observations and it provides evidence of absence of this association in a tropical country that does not experience marked seasons.

Keywords: suicide; weather; seasons; mental health; Colombia

---

1. Introduction

Suicide is an important global public health problem that has been increasing in the last three decades [1]. Current estimates state that 75% of all suicides occur in low- and medium-income countries [2]. Suicide is a multifactorial event with diverse biological and sociocultural determinants, such as mental health, social support networks, physical health, and significant life events [3]. In addition, previous studies have suggested that seasonal changes in suicide rate could be associated with changes in weather, although, the majority of this data comes from developed countries outside of the tropics where changes in weather conditions are more pronounced within the span of a year [3,4].

In the last two decades, more advanced statistical analyses and improved information systems have helped to confirm peaks in suicide in countries that experience seasons, especially during the spring and summer periods [5,6]. In particular, it was suggested that these peaks are related to changes in temperature [7–9], which in turn, would subsequently cause an increase in the levels of serotonin

(5-hydroxytryptamine or 5-HT) and lead to states of greater impulsiveness and aggressiveness during periods of hot weather [9]. However, these findings were based on divergent evidence in which peaks were also associated with low temperatures [10]. Moreover, some other studies found no association with any of the weather variables [11]. There are also contradictory studies regarding the relationship between suicide incidence and precipitation and relative humidity, albeit based on less evidence [8,12,13].

Changes in season and weather conditions usually correspond to periods of the year in which there are also changes in social activities, such as summer holidays and certain festivities; it is methodologically challenging to separate the potential effects of the two types of determinants [7]. For example, in countries that experience marked seasons, such as Italy, an increase in suicides has been reported during December and April, months that coincide with Christmas and Easter, respectively [13–15]. Similar results in which special dates are related with suicide have also been observed in developing countries, such as Mexico [16]. The changes in social dynamics that occur during such dates has been proposed as the underlying factor of this phenomenon [17]. In fact, psychopathological symptoms tend to increase during holidays as a result of the changes in social interactions and the disappointment of unmet psychological expectations [18–20]. For this reason, any study analyzing the potential effects of weather on suicide should consider holidays as a possible confounding factor in the analysis.

Literature dealing with the link between weather and suicide is very limited for medium-income countries that do not experience seasons. This is true for Colombia, which apart from having a different sociocultural context, experiences less fluctuations in weather when compared to countries that are located outside the tropics. Accordingly, one hypothesis to be explored is that if there is a true association between weather variables and suicides, this association should be also present in cities with lower weather variance, and remain even after adjusting by festivities. Therefore, evaluating the association between weather and suicide in Colombia will deepen our understanding of this relationship and help to establish a more global hypothesis of the causal relationships behind this phenomenon.

Here, our objective was to estimate the association between weather variables (temperature and rainfall) and the daily incidence of suicide among men and women living in the five major cities of Colombia between 2005 and 2015.

2. Materials and Methods

2.1. Study Location

Colombia is located in the northwest region of South America, and is approximately equidistant from the two extremities of the American continent. It extends from 4°13′30″ S to 12°27′46″ N, and from 66°50′54″ W to 79°0′23″ W. With a total land area of 1,141,748 km², Colombia is the fourth largest country of South America. As Colombia is located in the intertropical zone, it has a large variety of weathers and ecosystems. Having an equatorial location (the equator line crosses the south part of the country), Colombia does not have drastic changes of sunlight all year long, and seasonal variations are related only to rainy and dry periods.

The temperature in Colombia remains stable all year long; therefore, differences across cities are principally attributed to differences in geographic location (altitude). These differences are associated with variations in vegetation across the entire country. As such, the country is divided into six geographical regions: the Andean, Caribbean, Pacific, Orinoco, Amazon, and Insular regions. Bogotá (D.C.), Medellin, and Bucaramanga are located in the Andean region, but due to their different altitudes, they experience fairly different weathers. Whereas, Bogotá has an annual average temperature of 14 °C, Medellin and Bucaramanga averages are higher, 22 °C and 23 °C, respectively. On the other hand, Barranquilla and Cali are located in the Caribbean and Pacific regions, respectively. While Barranquilla is located at an altitude of 18 m and has annual average temperature of 27 °C, Cali has values of 1018 m and 24 °C. Having these heterogeneous weather conditions between cities, there are no strong variations in temperature within cities over the year.

2.2. Study Design

We conducted a multi-temporal, multi-city ecological study. The observational unit was the day by city from the 1st of January 2005 to the 31st of December 2015. We included the five capital cities of Colombia with the largest population (Bogotá, Medellin, Cali, Barranquilla, and Bucaramanga). In 2015, the population of these five cities accounted for approximately 30% of the total population of Colombia and 39.2% of the total urban population.

2.3. Variables and Data Sources

2.3.1. Response Variables: Suicide Counts

For each city, we determined the number of suicides per day by counting the number of deaths whose cause had been classified as one of the International Classification of Diseases (ICD)-10 codes X60-X84 or Y87.0. This information was obtained from the mortality database of the National Department of Statistics (DANE) for the whole period of the study. The total population by city and year was obtained from the DANE population projections, and was used as exposure variable in the analyses. By assuming that the total population of a city does not change within the course of a year, we were able to compare the suicide counts between cities.

2.3.2. Independent Variables: Weather Variables

We obtained daily data for temperature in degrees Celsius (average) and rainfall in millimeters (total amount). For the whole period 2005–2015, these data were obtained from the official historical registry of the meteorological station at each city local airport; therefore, information was obtained from one meteorological station by city. The daily average was computed as the arithmetic mean of hourly data when 75% or more of hourly data were available. Overall, information was available for 92% and 91% of the days for temperature and rainfall, respectively.

In addition, we considered holidays to be a principal confounding factor. As such, we generated an indicator variable for holidays that consisted of the following four categories: regular working day (reference category), weekday holiday, long weekend, and special day but without the day off. Given that these days change every year in Colombia, we used official historical calendars to manually recover this information.

2.4. Statistical Analyses

All of the weather variables and suicide counts were summarized using measures of central tendency (mean and median) and dispersion (standard deviation and interquartile range). The distribution of the daily suicide counts among men and women was assessed using the dispersion index test (VIT) [21] and the Bohning asymptotic test [22]. As this test rejected the null hypothesis of equidispersion, we reasonably supposed a Poisson distribution of suicides for our analyses. Additionally, the autocorrelation in suicides rates was verified for both sexes using partial autocorrelation graphs and the Wallis test. We did not reject the null hypothesis in all of the mentioned tests ($p > 0.10$), so we did not have evidence of a seasonal trend in suicide rates.

We use multi-city conditional Poisson models for the number of daily suicides using the yearly population (total population and population by sex) as the exposure variable, the weather parameters as the main independent variables, and holidays as the main confounding factor. We used Poisson models that were conditioned by time strata (grouping by day, month, and year) to control for the seasonality of suicide data. In these models, weather effects are estimated when considering the structure of the correlation that the observations would have on the same stratum of day of the week, month, and year [23], and considering the city as a fixed effect.

We did not find evidence of a non-linear relationship between any of the independent variables and the logarithm of the expected value of the suicides rates when these relationships were explored graphically or using quartiles of the weather variables. Therefore, we used the independent variables in a continuous scale.

We used multiplicative terms between each independent variable and the city indicator to explore heterogeneity between cities. This enabled us to obtain adjusted estimates of the coefficients for each city, estimates that are equivalent to performing city-stratified analyses, but more efficient.

Subsequently, as an additional analysis, by using the moving averages of temperature and rainfall over the last seven days in each city as the main independent variables, we examined potential lagged effects, explored from one up to seven days before each observation, for each independent variable. Finally, we created fixed Poisson models that explored the cumulative effect of each variable.

All of the adjusted models were stratified by sex, and all of the assumptions were verified. An association was considered to be statistically significant at an alpha of 0.05. All of the analyses were performed using STATA 12 (Stata Corporation, College Station, TX, USA).

3. Results

As a large quantity of the days in our study period were in fact, suicide-free days, we presented the suicide trend in men and women as a function of consecutive weeks (Figure 1). The average number of suicides per day for men and women, respectively, was 0.57 and 0.14 in Bogotá, 0.28 and 0.07 in Medellin, 0.21 and 0.04 in Cali, 0.06 and 0.02 in Bucaramanga, and finally, 0.09 and 0.01 in Barranquilla.

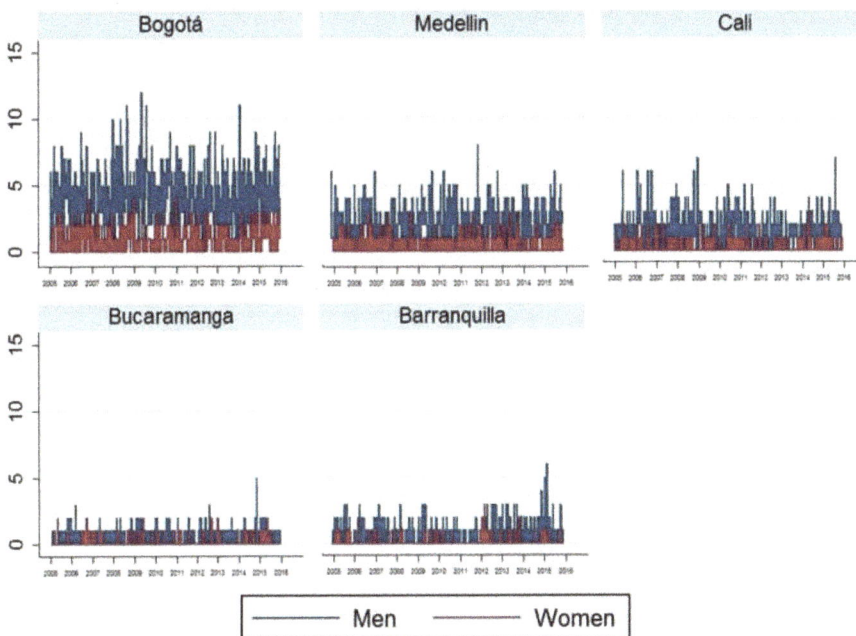

Figure 1. Weekly suicide count for the five main cities of Colombia, 2005–2015.

Figure 2 displays the time series of the average daily temperature and rainfall for each of the five cities during the study period. This data, along with its central tendency and dispersion measures, is summarized in Table 1. Bogotá presents the lowest average temperature (13.67 °C), with a minimum value of 8.20 °C and a maximum of 18.20 °C. With respect to rainfall, Cali has the highest average value (3.98 mm), with a maximum of 185.66 mm.

Table 1. Distribution of the daily weather variables for the five main cities of Colombia.

	Mean	sd	p25	p50	p75	min	max
Bogotá							
Temperature	13.67	0.94	13.10	13.70	14.30	8.20	18.20
Rainfall	2.19	5.17	0.00	0.06	2.03	0.00	115.06
Medellin							
Temperature	23.49	1.41	22.60	23.44	24.30	18.10	31.60
Rainfall	3.30	11.19	0.00	0.00	2.03	0.00	229.62
Cali							
Temperature	24.07	1.38	23.10	24.07	25.10	19.90	29.00
Rainfall	3.98	13.01	0.00	0.00	0.89	0.00	185.66
Bucaramanga							
Temperature	22.71	1.33	21.80	22.70	23.57	17.80	31.60
Rainfall	2.17	6.82	0.00	0.00	0.93	0.00	95.50
Barranquilla							
Temperature	27.91	1.06	27.20	27.90	28.70	24.30	32.90
Rainfall	1.56	6.62	0.00	0.00	0.00	0.00	105.92

sd: standar deviation, p: percentile

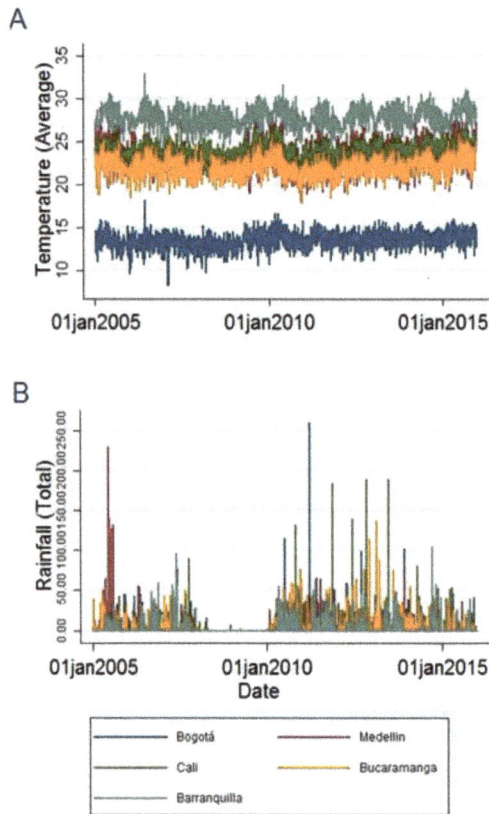

Figure 2. Weather variables of the five main cities of Colombia, 2005–2015. (**A**) Average daily temperature, (**B**) Daily rainfall.

The conditional Poisson models for suicide in men and women showed that temperature and rainfall do not present statistically significant associations with the number of suicides, neither in the general nor the sex-stratified models (Table 2). In contrast, with respect to the holiday indicator variable, we found that long weekends—in comparison to regular working days—not only have a statistically significant association with total suicides (Incidence Rate Ratio (IRR): 1.19, 95% confidence interval (CI): 1.04–1.23), but also with suicides in both sexes (IRR: 1.12, 95% CI: 0.97–1.31 for men and IRR: 1.51, 95% CI: 1.14–2.00 for women). However, this association is not observed for other categories of the same variable (i.e., weekday holiday or special day but without the day off).

Table 2. Conditional Poisson models for daily suicides among men and women in the five main Colombian cities, 2005–2015.

Variables	Men			Women			All		
	IRR	95% CI	*p*	IRR	95% CI	*p*	IRR	CI 95%	*p*
Temperature	1.01	0.97–1.05	0.51	1.06	0.98–1.16	0.15	1.02	0.99–1.06	0.21
Rainfall	1.00	1.00–1.00	0.61	1.00	0.99–1.01	0.73	1.00	1.00–1.00	0.74
Special dates (Ref: Regular working day)									
Weekday holiday	1.04	0.88–1.24	0.62	1.12	0.79–1.58	0.52	1.05	0.91–1.23	0.49
Long weekend	1.12	0.97–1.31	0.13	1.51	1.14–2.00	<0.01	1.19	1.04–1.23	0.01
Special day without the day off	1.10	0.81–1.51	0.54	1.49	0.83–2.67	2.67	1.17	0.88–1.54	0.27
*Temperature (deviation from mean)	1.01	0.97–1.06	0.55	1.08	0.98–1.18	0.11	1.03	0.99–1.07	0.21

* Estimations obtained from independent models using deviation of historical average temperature by city. In these models, other coefficients did not change. IRR = Incidence Rate Ratio. CI: confidence interval.

Figure 3 shows the graphical results of the Poisson models by city while considering the heterogeneity of each weather variable. For each of the five cities, we found that the null hypothesis could not be rejected for any of the analyzed weather variables. These results justify the use of pooled models (such as the ones that are proposed in this study) and they rule out the possibility that an association could have existed in only some of the cities.

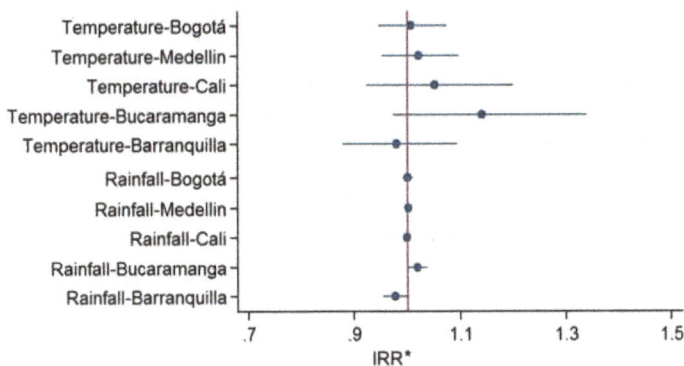

Figure 3. Estimators obtained for the weather variables in the five main cities of Colombia. Estimations were obtained using fixed-effect Poisson models and adjusted for the effect of holidays. *IRR can be interpreted as the average change in the suicide rate for each increment of one unit un each independent variable (a degree in the case if temperature, a cubic millimeter in the case of rainfall).

Finally, we did not find a statistically significant association between suicide counts and weather variables when considering their lagged (Table 3) and cumulative effects (Table 4) (i.e., the seven-day moving average).

Table 3. Fixed Poisson models for the lagged effects of weather variables on daily suicides among men and women in five Colombian cities, 2005–2015.

		Men				Women				All			
	Lag	IRR	95% CI		*p*	IRR	95% CI		*p*	IRR	95% CI		*p*
Temperature	L0	0.99	0.92	1.07	0.84	1.05	0.88	1.25	0.58	1.00	0.93	1.08	0.92
	L1	1.06	0.96	1.16	0.26	0.94	0.77	1.14	0.52	1.03	0.95	1.13	0.43
	L2	0.99	0.90	1.08	0.76	1.10	0.89	1.35	0.37	1.00	0.92	1.09	0.92
	L3	0.96	0.87	1.06	0.40	1.14	0.93	1.40	0.21	0.99	0.91	1.08	0.83
	L4	1.05	0.96	1.15	0.31	0.88	0.72	1.08	0.22	1.02	0.94	1.11	0.66
	L5	1.02	0.93	1.12	0.67	1.11	0.91	1.36	0.29	1.04	0.95	1.13	0.41
	L6	0.93	0.85	1.02	0.11	0.99	0.81	1.20	0.88	0.94	0.86	1.02	0.13
	L7	1.03	0.95	1.12	0.45	0.84	0.71	1.00	0.05	1.00	0.93	1.07	0.90
Rainfall	L0	1.00	0.98	1.01	0.56	1.00	0.98	1.03	0.85	1.00	0.99	1.01	0.71
	L1	1.01	0.99	1.02	0.47	1.00	0.97	1.04	0.87	1.01	0.99	1.02	0.44
	L2	1.00	0.99	1.02	0.63	1.01	0.97	1.04	0.69	1.00	0.99	1.02	0.53
	L3	0.99	0.98	1.01	0.36	1.03	1.00	1.07	0.05	1.00	0.99	1.01	0.95
	L4	1.01	0.99	1.02	0.32	0.98	0.95	1.01	0.20	1.00	0.99	1.02	0.70
	L5	0.99	0.98	1.01	0.51	1.00	0.97	1.04	0.88	1.00	0.98	1.01	0.60
	L6	1.00	0.98	1.01	0.89	1.00	0.97	1.03	0.89	1.00	0.99	1.01	0.89
	L7	1.00	0.99	1.02	0.77	0.97	0.95	1.00	0.06	1.00	0.99	1.01	0.66

Table 4. Conditional Poisson models for the cumulative effects of weather variables on daily suicides among men and women in five Colombian cities, 2005–2015

	Men			Women			All		
	IRR	95% CI	*p*	IRR	95% CI	*p*	IRR	95% CI	*p*
Temperature	1.02	0.98–1.07	0.25	1.04	0.95–1.14	0.38	1.03	0.93–1.07	0.16
Rainfall	1.00	0.99–1.00	0.13	1.00	0.99–1.01	0.21	0.99	0.99–1.00	0.12

Estimations have been adjusted for the effect of holidays. The estimators represent changes with respect to the moving average of each weather variable in the last 7 days.

4. Discussion

Here, we found no evidence supporting an association between weather (temperature and rainfall) and the daily suicide incidence among men and women of the five major cities of Colombia. However, we did find a statistically significant association between daily suicide counts and holidays, specifically long weekends. Furthermore, our results show that the association between suicide and holidays persists, even in the absence of marked variations in weather. This finding supports the fact that holidays are an important confounding factor to consider when studying the association between suicide and weather in any context. Controlling for holidays allows for us to focus on the potential explanatory mechanisms of the association between weather and suicide, such as geographic and cultural processes, more than just the associated physiological changes in the human body, as previous studies linking temperature effects on mental health have proposed [24,25]. The lack of evidence of a significant association between suicide and weather in this study might be explained by the fact that this relationship is modified in countries without drastic seasonal changes of weather, such as Colombia [26], or by potential confounding bias in previous studies.

As there is currently a scarce amount of information available regarding this association in tropical or equatorial countries, our study brings to light important results. One study that analyzed this association in Sao Paulo—a city that is located in the southeast of Brazil close to the Tropic of Capricorn—found that variables, such as hours of sunlight and average temperature, had no effect on suicide during the 1996–2004 time period (based on official suicide registries) [27]. Another study conducted in Singapore [28]—an Asian country that is located one-degree north of the equator—neither found an association between maximum temperature and suicides when analyzing data from 1980 to 1989. The only exception to this was among Malaysian adolescents, for whom high temperatures appeared to account for about 58% of the variation in suicide incidence. It is also interesting to point out that upon stratifying the analyses, suicides committed by the Hindu population presented peaks in the months of April, September, and November, as well as during Deepavali, or the festival of the lights, which is the most important festival for the Hindu population, usually celebrated in the month of November. While the authors suggest that increased tensions due to extra expenses and forced family gatherings could be related to the rise in suicide during this festival, they also acknowledge that the existence of a true sociocultural association is not very clear [28].

The association between suicide and holidays has already been reported in studies that are conducted in Colombia. For example, while in the 2000–2010 time period there were a total of 24,882 suicides, the average daily number increased from an average of 6.2 to 8.0 when considering holidays. Furthermore, the largest number of suicides during this period was found to occur on the 1st of January and the 25th of December [29]. In fact, it is well known that psychopathological symptoms tend to increase around the holidays, including Christmas [18,19]. Such symptoms are enhanced during the holidays because during this time susceptible individuals experience an increase in failed social encounters, and consequently feel more frustrated, a feeling that is also consistent with an increase in dysphoric moods [20]. This susceptibility might be relevant, especially in people with a history of psychiatric disorders, as reported in a meta-analysis of 3275 suicides, where 87.3% of suicide cases had a history of psychiatric disorders [30].

An increase in the number of suicides during the holidays can also be the result of other causes, especially when considering that holidays are periods with a wide variety of stressors, including higher levels of alcohol consumption, changes in sleep rhythms, increased economic pressure, and more family conflicts [15]. Indeed, it has been reported that alcohol and drug consumption is related to 25–50% of all suicides [31], and that this risk is increased even further when the consumption is comorbid with other mental disorders [1].

These associations have also been widely explained by the broken promise effect of Gabennesh [32], in which spring, the weekends, and holidays are typically seen as positive events that can sometimes hold false promises (i.e., they are seen to offer more than what they can actually deliver). In such a way, these events can lead to a sense of hope and create expectations that are not met. This type of situation could be aggravated by a lack of social contacts [14] and/or by a more limited access to care and help [33], a situation that often occurs during holidays when services are closed or have reduced working hours [15]. Therefore, adding this negative emotional impact to the lack of social support could be a relevant risk factor, and have a greater impact in people with some affective disorder.

From the above discussion, it becomes evident that, when analyzing the relationship between suicide and weather variables, it is also important to take into account and isolate the effects of holidays and seasons. As holiday effect might interact with the weather variables, it becomes extremely difficult to distinguish between the two effects [12]. In our study, however, the geographical location of Colombia not only made it possible to isolate the seasonality effect, but also to explore the association in cities that are subjected to different weather [34].

We found no association between suicide and weather variables using data from the largest Colombian cities that helps to clarify this complex, heterogenic association in a country that does not experience marked seasonal variations. Regarding the quality of the suicide data, it is important to note that the data was obtained from the National Mortality Registry from the DANE. A quality assessment of the mortality registry was conducted by Cendales y Pardo [35] for the death certificates between 2002 and 2006 in Colombia and their results showed that 92.8% of the deaths were certified and coded correctly; the authors concluded with evidence that National Mortality Registry has a good quality in Colombia. Moreover, the underregistration of suicide seems to not be a problem in Colombia as previous studies have confirmed a steadily increase in the suicide registries, especially since 1998 [36]. In terms of weather data, the meteorological variables data was obtained from official reports for meteorological stations at local airports in all cities. Meteorological stations and registries are regulated by the Aeronáutica Civil de Colombia and follow quality system guideless, which support the quality of the registries.

Therefore, our results provide important evidence that can be generalizable to other countries that are located close to the equator; however, more research in this geographical context is necessary to confirm the consistency of findings. Different from the countries where this association has been studied extensively, countries in the tropics only have dry and rainfall seasons. In the particular case of Colombia, the subdivision of the Andes mountain range in the national territory causes the formation of local and regional climates of high complexity and difficulty of prediction at different moments of time.

The main limitation of our study is its ecological design. As this type of study does not include individual variables, conclusions were interpreted for the population as a whole and cannot be extrapolated to the individual level. Another important limitation is that suicide records could be prone to errors at some stage of production, including collection, coding, data processing, or diagnosis [37]. This, in turn, would lead to the possibility of misclassified individuals. This type of scenario can lead to differential measurement errors. In addition, the small number of suicides is often a problem in these studies. Although we used data from the five largest Colombian cites, there is still a probability that type 2 error could explain the absence of any significant association. In this way, the low power of this study could be explained by the fact that Colombia, comparatively with other countries of the region, has a low suicide incidence [38], which could make difficult the finding of potential associations, especially those of low magnitude.

Finally, we use a very limited number of weather variables. In addition to the two weather variables that are evaluated here, the absolute humidity, the amount of sunlight radiation (Mj/m^2) [39], and barometric pressure [24], were not available from the source data and should also be considered as potential variables that influence suicide. However, previous studies assessing weather and suicide have identified temperature and rainfall as the main weather factors that are associated with suicides [27,28], and our study included cities with different mean temperatures in the analysis. On the other hand, data concerning the exact location of a suicide (e.g., inside or outside of the home) should also be taken into account as this type of information clarifies whether weather conditions, such as high or low temperatures and rainfall, could have been a direct influencing factor [40].

In the future, more studies should be conducted to examine this association in countries that are located close to the equator or the tropics. By analyzing other counties that have different cultures and holidays, it would be possible to isolate the effect of holidays in the analysis. In addition, by considering risky behaviors that occur on holidays, whether a suicide takes place inside or outside of home, relevant social determinants, history, or psychiatric disorders, and the association with other elements of suicidal behavior, such as ideation, planning, or attempt, it would be possible to clarify further if the climatic variables are truly associated with suicide independently.

Finally, it would be also important to replicate studies conducted in different countries using similar methodologies, as the heterogeneity of the associations reported in other latitudes could be related to the use of different time frames, units of time (e.g., day or month), statistical methods, geographic locations, and socioeconomic conditions [24,26,37].

5. Conclusions

Overall, our results suggest that there is no association between weather variables and the incidence of suicide, at least in the five largest Colombian cities. However, we did find an association between suicide and holidays, specifically long weekends. As this association could be related to behavioral patterns and opportunities, it is crucial that prevention programs start to consider these elements, especially in the identification of high-risk groups (such as people with history of psychiatric disorders, mental health comorbidities, and risk behaviors, such as alcohol consumption) and promoting a monitoring of these factors, especially during holidays.

Author Contributions: J.A.F.-N. and L.A.R.-V.: conception and design, J.A.F.-N., L.A.R.-V., V.A.F.-G. and C.I.A.-G.: acquisition of data, analysis, and interpretation of data; J.A.F.-N., L.A.R.-V. and C.I.A.-G.: drafting the article and revising it critically for important intellectual content. All authors read and approved the final manuscript.

Funding: This research received no external funding.

Conflicts of Interest: The authors declare no conflict of interest.

References

1. WHO. *Preventing Suicide: A Global Imperative*; World Health Organization: Washington, DC, USA, 2014.
2. Fleischmann, A.; Arensman, E.; Berman, A.; Carli, V.; De Leo, D.; Hadlaczky, G.; Howlader, S.; Vijayakumar, L.; Wasserman, D.; Saxena, S. Overview evidence on interventions for population suicide with an eye to identifying best-supported strategies for LMICs. *Glob. Ment. Health* **2016**, *3*, e5. [CrossRef] [PubMed]
3. Cash, S.J.; Bridge, J.A. Epidemiology of youth suicide and suicidal behavior. *Curr. Opin. Pediatr.* **2009**, *21*, 613–619. [CrossRef] [PubMed]
4. Amitai, M.; Apter, A. Social Aspects of Suicidal Behavior and Prevention in Early Life: A Review. *Int. J. Environ. Res. Public Health* **2012**, *9*, 985–994. [CrossRef] [PubMed]
5. Tsai, J.-F.; Cho, W. Temperature change dominates the suicidal seasonality in Taiwan: A time-series analysis. *J. Affect. Disord.* **2012**, *136*, 412–418. [CrossRef] [PubMed]
6. Likhvar, V.; Honda, Y.; Ono, M. Relation between temperature and suicide mortality in Japan in the presence of other confounding factors using time-series analysis with a semiparametric approach. *Environ. Health Prev. Med.* **2011**, *16*, 36–43. [CrossRef] [PubMed]
7. Fernández-Arteaga, V.; Tovilla-Zárate, C.A.; Fresán, A.; González-Castro, T.B.; Juárez-Rojop, I.E.; López-Narváez, L.; Hernández-Díaz, Y. Association between completed suicide and environmental temperature in a Mexican population, using the Knowledge Discovery in Database approach. *Comput. Methods Programs Biomed.* **2016**, *135*, 219–224. [CrossRef] [PubMed]
8. Tsai, J.-F. Socioeconomic factors outweigh climate in the regional difference of suicide death rate in Taiwan. *Psychiatry Res.* **2010**, *179*, 212–216. [CrossRef] [PubMed]
9. Qi, X.; Hu, W.; Page, A.; Tong, S. Associations between climate variability, unemployment and suicide in Australia: A multicity study. *BMC Psychiatry* **2015**, *15*, 114. [CrossRef] [PubMed]
10. Maes, M. Seasonal Variation in Plasma L-Tryptophan Availability in Healthy Volunteers. *Arch. Gen. Psychiatry* **1995**, *52*, 937–946. [CrossRef] [PubMed]
11. Grove, O.; Lynge, J. Suicide and attempted suicide in Greenland. A controlled study in Nuuk (Godthaab). *Acta Psychiatr. Scand.* **1979**, *60*, 375–391. [CrossRef] [PubMed]
12. Deisenhammer, E.A. Weather and suicide: The present state of knowledge on the association of meteorological factors with suicidal behaviour. *Acta Psychiatr. Scand.* **2003**, *108*, 402–409. [CrossRef] [PubMed]
13. Preti, A.; Miotto, P. Seasonality in suicides: The influence of suicide method, gender and age on suicide distribution in Italy. *Psychiatry Res.* **1998**, *81*, 219–231. [CrossRef]
14. Jessen, G.; Jensen, B.F.; Arensman, E.; Bille-Brahe, U.; Crepet, P.; Leo, D.D.; Hawton, K.; Haring, C.; Hjelmeland, H.; Michel, K.; et al. Attempted suicide and major public holidays in Europe: Findings from the WHO/EURO Multicentre Study on Parasuicide. *Acta Psychiatr. Scand.* **1999**, *99*, 412–418. [CrossRef] [PubMed]
15. Jessen, G.; Jensen, B.F. Postponed suicide death? Suicides around birthdays and major public holidays. *Suicide Life-Threat. Behav.* **1999**, *29*, 272–283. [PubMed]
16. Fernández-Niño, J.A.; Astudillo-García, C.I.; Bojorquez-Chapela, I.; Morales-Carmona, E.; Montoya-Rodriguez, A.A.; Palacio-Mejia, L.S. The Mexican Cycle of Suicide: A National Analysis of Seasonality, 2000–2013. *PLoS ONE* **2016**, *11*, e0146495. [CrossRef] [PubMed]
17. Durkheim, E. *Suicide: A Study in Sociology (JA Spaulding & G. Simpson, Trans.)*; Free Press: Glencoe, IL, USA, 1951.
18. Sansone, R.A.; Sansone, L.A. The christmas effect on psychopathology. *Innov. Clin. Neurosci.* **2011**, *8*, 10–13. [PubMed]
19. Plöderl, M.; Fartacek, C.; Kunrath, S.; Pichler, E.-M.; Fartacek, R.; Datz, C.; Niederseer, D. Nothing like Christmas—Suicides during Christmas and other holidays in Austria. *Eur. J. Public Health* **2015**, *25*, 410–413. [CrossRef] [PubMed]
20. Friedberg, R.D. Holidays and emotional distress: Not the villains they are perceived to be. *Psychology* **1990–1991**, *27–28*, 59–61.

21. Cochran, W.G. Some Methods for Strengthening the Common χ^2 Tests. *Biometrics* **1954**, *10*, 417–451. [CrossRef]

22. Bohning, D. A Note on a Test for Poisson Overdispersion. *Biometrika* **1994**, *81*, 418–419. [CrossRef]

23. Armstrong, B.G.; Gasparrini, A.; Tobias, A. Conditional Poisson models: A flexible alternative to conditional logistic case cross-over analysis. *BMC Med. Res. Methodol.* **2014**, *14*, 122. [CrossRef] [PubMed]

24. Dixon, P.G.; Sinyor, M.; Schaffer, A.; Levitt, A.; Haney, C.R.; Ellis, K.N.; Sheridan, S.C. Association of Weekly Suicide Rates with Temperature Anomalies in Two Different Climate Types. *Int. J. Environ. Res. Public Health* **2014**, *11*, 11627–11644. [CrossRef] [PubMed]

25. Dixon, K.W.; Shulman, M.D. A statistical investigation into the relationship between meteorological parameters and suicide. *Int. J. Biometeorol.* **1983**, *27*, 93–105. [CrossRef] [PubMed]

26. Dixon, P.G.; Kalkstein, A.J. Where are weather-suicide associations valid? An examination of nine US counties with varying seasonality. *Int. J. Biometeorol.* **2018**, *62*, 685–697. [CrossRef] [PubMed]

27. Nejar, K.A.; Bensenor, I.M.; Lotufo, P.A. Sunshine and suicide at the tropic of Capricorn, Sao Paulo, Brazil, 1996–2004. *Rev. Saude Pública* **2007**, *41*, 1062–1064. [CrossRef] [PubMed]

28. Kok, L.P.; Tsoi, W.F. Season, climate and suicide in Singapore. *Med. Sci. Law* **1993**, *33*, 247–252. [CrossRef] [PubMed]

29. Cardona, D.; Medina-Pérez, O.; Cardona, D. Caracterización del suicidio en Colombia, 2000–2010. *Rev. Colomb. Psiquiatr.* **2016**, *45*, 170–177. [CrossRef] [PubMed]

30. Arsenault-Lapierre, G.; Kim, C.; Turecki, G. Psychiatric diagnoses in 3275 suicides: A meta-analysis. *BMC Psychiatry* **2004**, *4*, 37. [CrossRef] [PubMed]

31. Schneider, B. Substance use disorders and risk for completed suicide. *Arch. Suicide Res.* **2009**, *13*, 303–316. [CrossRef] [PubMed]

32. Gabennesch, H. When Promises Fail: A Theory of Temporal Fluctuations in Suicide. *Soc. Forces* **1988**, *67*, 129–145. [CrossRef]

33. Hadlaczky, G.; Hokby, S. Increased suicides during new year, but not during Christmas in Sweden: Analysis of cause of death data 2006–2015. *Nordic J. Psychiatry* **2018**, *72*, 72–74. [CrossRef] [PubMed]

34. Ajdacic-Gross, V.; Lauber, C.; Sansossio, R.; Bopp, M.; Eich, D.; Gostynski, M.; Gutzwiller, F.; Rössler, W. Seasonal associations between weather conditions and suicide—Evidence against a classic hypothesis. *Am. J. Epidemiol.* **2007**, *165*, 561–569. [CrossRef] [PubMed]

35. Cendales, R.; Pardo, C. Colombian death certificate quality, 2002–2006. *Rev. Salud Pública* **2011**, *13*, 229–238. [CrossRef] [PubMed]

36. Cendales, R.; Vanegas, C.; Fierro, M.; Córdoba, R.; Olarte, A. Suicide trends in Colombia, 1985–2002. *Rev. Panam. Salud Pública* **2007**, *22*, 231–238. [CrossRef] [PubMed]

37. OPS. *Sobre la Estimación de Tasas de Mortalidad Para Países de la Región de las Américas*; Organización Panamericana de la Salud: Washington, DC, USA, 2003.

38. Mascayano, F.; Irrarázabal, M.; DEmilia, W.; Shah, B.; Varner, S.J.; Sapag, J.C.; Alvarado, R.; Hsin Yang, L. Suicidio en Latinoamérica: Un creciente problema de salud pública. *Rev. Fac. Cienc. Méd.* **2015**, *73*, 295–303.

39. Bando, D.H.; Teng, C.T.; Volpe, F.M.; Masi, E.; Pereira, L.A.; Braga, A.L. Suicide and meteorological factors in Sao Paulo, Brazil, 1996–2011: A time series analysis. *Rev. Bras. Psiquiatr.* **2017**, *39*, 220–227. [CrossRef] [PubMed]

40. Guo, Y.; Barnett, A.G. Invited commentary: Assessment of air pollution and suicide risk. *Am. J. Epidemiol.* **2015**, *181*, 304–308. [CrossRef] [PubMed]

International Journal of
*Environmental Research
and Public Health*

MDPI

Article

An Ecological Study on the Spatially Varying Relationship between County-Level Suicide Rates and Altitude in the United States

Hoehun Ha [1],* and Wei Tu [2]

[1]   Department of Sociology, Anthropology and Geography, Auburn University at Montgomery,
      7041 Senators Drive, Montgomery, AL 36117, USA
[2]   Department of Geology and Geography, Georgia Southern University, P.O. Box 8149, Statesboro,
      GA 30460, USA; wtu@georgiasouthern.edu
*   Correspondence: hha@aum.edu; Tel.: +1-334-244-3291

Received: 8 February 2018; Accepted: 1 April 2018; Published: 4 April 2018

Abstract: Suicide is a serious but preventable public health issue. Several previous studies have revealed a positive association between altitude and suicide rates at the county level in the contiguous United States. We assessed the association between suicide rates and altitude using a cross-county ecological study design. Data on suicide rates were obtained from a Web-based Injury Statistics Query and Reporting System (WISQARS), maintained by the U.S. National Center for Injury Prevention and Control (NCIPC). Altitude data were collected from the United States Geological Survey (USGS). We employed an ordinary least square (OLS) regression to model the association between altitude and suicide rates in 3064 counties in the contiguous U.S. We conducted a geographically weighted regression (GWR) to examine the spatially varying relationship between suicide rates and altitude after controlling for several well-established covariates. A significant positive association between altitude and suicide rates (average county rates between 2008 and 2014) was found in the dataset in the OLS model ($R^2 = 0.483$, $p < 0.001$). Our GWR model fitted the data better, as indicated by an improved R^2 (average: 0.62; range: 0.21–0.64) and a lower Akaike Information Criteria (AIC) value (13,593.68 vs. 14,432.14 in the OLS model). The GWR model also significantly reduced the spatial autocorrelation, as indicated by Moran's I test statistic (Moran's $I = 0.171$; z = 33.656; $p < 0.001$ vs. Moran's $I = 0.323$; z = 63.526; $p < 0.001$ in the OLS model). In addition, a stronger positive relationship was detected in areas of the northern regions, northern plain regions, and southeastern regions in the U.S. Our study confirmed a varying overall positive relationship between altitude and suicide. Future research may consider controlling more predictor variables in regression models, such as firearm ownership, religion, and access to mental health services.

Keywords: altitude; suicide rates; geographically weighted regression; health geography

---

1. Introduction

Suicide, increasingly recognized as a serious but preventable public health problem, is an important cause of global mortality [1,2]. Suicide claimed close to 800,000 lives worldwide in 2015, accounting for 1.4% of the total of all-cause deaths [3]. In the U.S. in 2015, there were more than 44,000 reported suicide deaths, making it the 2nd and 10th leading cause of death for the population aged between 15 and 34 and for all age groups, respectively. In addition, 9.8 million adults aged 18 or older had serious suicidal thoughts, and 1.4 million attempted suicide. The economic burden of suicide in 2013 alone was estimated to be between US$50.8 billion [4] and $93.5 billion [5]. Moreover, suicide rates increased from 10.5 per 100,000 in 1999 to 13.0 per 100,000 in 2015, a 24% growth in less than 20 years [6].

Past studies have implicated several risk factors for suicide, including demographic (e.g., certain age cohorts, white males, marital status), socioeconomic (e.g., poverty, unemployment, social isolation), environmental (e.g., altitude, air pollution, meteorological conditions), and behavioral (e.g., smoking, substance abuse) [7–16]. Other possible suicide triggers include anxiety or depression, psychiatric illness, mood disorders, and sociocultural factors [17–19]. However, risk factors may be different across geographic regions, and risk factors may also change over time. For instance, political violence, an uncommon risk factor for suicide in adolescents, is reportedly associated with suicide among Palestinian youths [20] and periodical economic recession was also found to be associated with elevated suicide rates [21,22].

According to the U.S. Centers for Disease Control and Prevention (U.S. CDC), the age-adjusted average annual suicide rates in the western, southern, mid-western, and northeastern U.S. regions were 14.86, 11.45, 11.52 and 8.97, per 100,000 between 1981 and 1998, respectively. The rates in these four regions were 13.04, 12.37, 11.52 and 8.63 per 100,000 between 1999 and 2015, respectively [23] and the suicide rate in the western region has been significantly higher than its counterparts over both periods of time. One explanation for regional differences in suicide rates is that altitude is much higher in the western region. For instance, Lester (1980) examined state-level suicide rates (1959–1961) and the latitude and longitude of the state capitals in the 48 conterminous states. Lester found that the suicide rates were statistically higher in the western than in the eastern states [24]. Cheng et al. (2005) reported a strong positive correlation between the average suicide rates of counties where the state capitals were located, and mean county altitude (using the altitude or mean altitude of the weather stations in these counties), in the 48 contiguous U.S. states between 1979 and 1998 [25]. Brenner et al. (2011) confirmed a significant positive relationship between county-level suicide rates and the county-center's altitude, using the suicide rates data between 1979 and 1998 in the 48 contiguous U.S. states [17]. Kim et al. (2011) also analyzed the county-level suicide rates between 1979 and 1998 in the 48 contiguous U.S. states and reported a strong positive association between the average county altitude and both firearm-related and non-firearm-related suicide rates [26]. Moreover, Trgovac et al. (2015) found increased county-level male suicide rates in the western U.S., small areas throughout the Appalachian Region, and the Midwest, using data from 2000 to 2006. In addition, social isolation was theorized as the major cause behind the elevated male suicide risk in high altitude regions [27].

Past studies have also explored several possible pathways connecting suicide rates and altitude. From a social perspective, higher altitude regions may be at a higher suicide risk because these regions tend to be socially and culturally more isolated, and individuals with weak social connections are more likely to view suicide as a solution to personal problems [28,29]. From a social, political and cultural perspective, higher firearm ownership in the western U.S. may partially contribute to higher suicide rates in the region. Firearm suicide is not only the most common, but also the most lethal type of suicide. Since suicidal attempts tend to be impulsive, access to firearms at home will increase the chance of using firearms as a means of suicide [30–32]. From a biophysiological and psychological perspective, increased suicide rates could be attributed to depression and/or mood instability, related to high-altitude hypoxia [33–35]. However, systematic physiological responses to ambient oxygen levels are extremely complicated and more research is needed to establish a robust causal link between altitude, hypoxia, mood disorder and suicide risk [36,37].

Many previous studies have reported a consistent positive relationship between suicide rates and altitude in the U.S. after adjusting for known independent variables. Such a relationship was also observed elsewhere in the world, including Korea, Taiwan and Ecuador [25,38,39]. However, results in the literature have not always been consistent. For instance, no association was found in a study in Turkey that used suicide rates in 81 provinces in 2007 and 2008 [40]. In a Japanese study, the slope of habitable land was reported to be a more relevant predictor than altitude [41]. A field-based study conducted in the Himalayas and the Andes between 2009 and 2011 disproved the hypothesized link between altitude, depression and suicide, and indirectly rejected the association between suicide and altitude because the prevalence of depression was found to be low in elderly highlanders [42].

Furthermore, two methodological issues in the literature should be noted. First, correlation analysis, an exploratory data analysis tool and ordinary least square (OLS) regression, a global regression method, appeared to be the two most common analytical approaches [17,26,43]. OLS assumes a homogeneous relationship between the outcome variable and predictor variables (e.g., altitude) but a spatially stationary relationship often does not hold for health outcomes [44,45]. Geographically Weighted Regression (GWR), an extension of OLS regression, allows for the exploration of spatial non-stationarity through estimation of local, rather than global, parameters [46]. GWR has been applied extensively in health studies over the past 15 years [47–50] and has also been adopted in suicide studies [26,51]. However, to the best of our knowledge, GWR has yet to be used to analyze the relationship between suicide and altitude. Second, altitude in the previous studies was represented by a proxy variable such as the location (longitude and latitude) of the state capital [24], the altitude of the state capital city [43], the altitude of the county center [17], or a mean area altitude such as the state mean [52] or the county mean [24]. In addition, altitude was also modeled as an aggregated second-level variable in a multilevel regression model [53]. Among these approaches, the county mean altitude is so far the altitude estimation at the finest geographic scale for an ecological study where the same scale suicide data are also available.

In this study, we investigated the varying association between county-level suicide rates and the mean county altitude, using a more recent U.S. suicide data set (2008 to 2014). The primary objectives of this study were as follows: (1) to compare results from the OLS model and GWR model; (2) to examine whether, and to what extent, the relationship between suicide rates and altitude vary across the 48 contiguous U.S. states, and (3) to explore the association of suicide rates and altitude, after controlling for several county-level demographic, socioeconomic, and behavioral variables. We hope that the findings gleaned from this study can deepen our understanding of the spatial variation of suicide rates and inform policies and initiatives for more precise suicide prevention and intervention.

2. Materials and Methods

2.1. Suicide Rates

The suicide data were collected from the annual mortality files of the National Center for Health Statistics (NCHS) under the U.S. CDC throughout the Web-based Injury Statistics Query and Reporting System (WISQARS) [23,27]. WISQARS provides fatal injury data in all of the 3141 U.S. counties, of which 3102 are in the contiguous U.S. Age-adjusted smoothed suicide rates were downloaded and used in our study. The technical details about the smoothing and age adjusting can be found at WISQARS [23]. In addition, counties with less than 20 cumulative deaths were excluded from our analysis to maintain statistical stability of the data following the U.S. CDC standard [54]. Thus, suicide rates from a total of 3064 counties between 2008 and 2014 were retained and used in our analyses (Figure 1).

2.2. Mean County Altitude Data

A digital altitude model (DEM), with 100-m spatial resolution for the 3064 counties (or equivalent units), was collected from the United States Geologic Survey [55]. The mean altitude of each county was calculated using a zonal statistics operation using the ArcGIS 10.3 software package (Esri, Redlands, CA, United States) [56]. County boundaries were collected from the U.S. Census Bureau [57] (Figure 2).

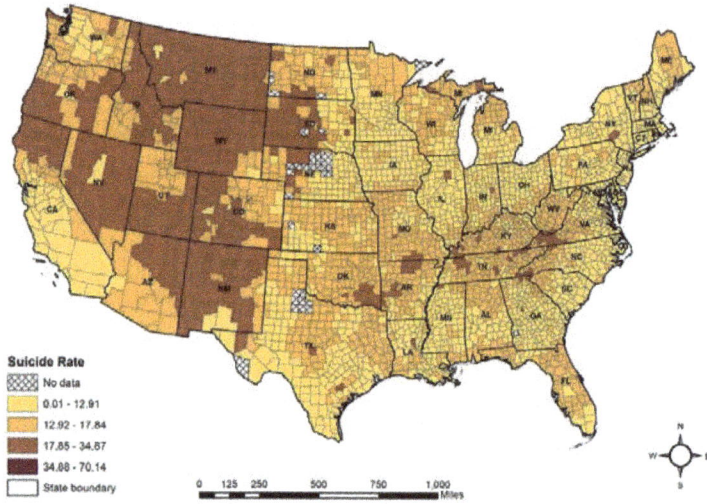

Figure 1. County-level average suicide rates in the 48 contiguous U.S. states between 2008 and 2014 (smoothed and age-adjusted).

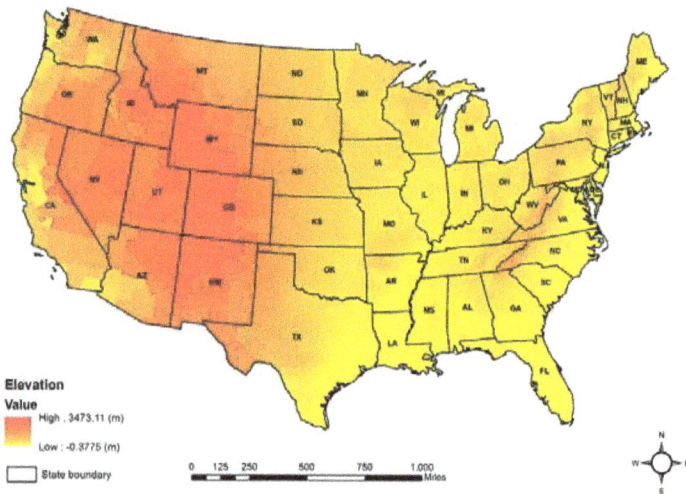

Figure 2. Mean county altitude in the 48 contiguous U.S. states.

2.3. Independent Variables

We accounted for the effects of potential confounding variables to enhance confidence levels in our analyses [58]. The variables were selected from four different categories based on the literature: (1) health behavior and clinical care variables; (2) social and economic variables; (3) physical environmental variables; and (4) demographic variables, as shown in Table 1 [7–16]. These county-level variables were collected from various data sources because variables possibly relevant to the suicide rate were not available and not sufficient from NCHS [58]. Moreover, for some predictor variables identified in the literature, such as firearm ownership, religious affiliation, and access to mental health services, county-level data were either unavailable or incomplete, so they were not included in our analyses.

Data of two demographic variables, percentage of the population aged 65 and above, and percentage of the African American population were obtained from the U.S. Census Population Estimates [59]. The health behavioral variable, percentage of the currently smoking population, was collected from the Selected Metropolitan/Micropolitan Area Risk Trends of Behavioral Risk Factor Surveillance System (SMART BRFSS). SMART BRFSS is a premier nationwide system of health-related telephone surveys in the U.S., collecting city- and county-level data about health-related risk behaviors and chronic health conditions [60,61]. In addition, the number of membership associations per 10,000 people (here after the association rate) was obtained from the American Community Survey (ACS) 5-year estimate (2010–2014). Specifically, associations include various membership organizations, such as civic organizations, sports organizations, political organizations, labor organizations, business organizations, and professional organizations. The association rate was used as a proxy indicator to quantify the degree of social support [62]. Finally, four predictor variables, selected from thirteen independent variables, were kept in our final regression models according to the results of statistical testing (Table 2). In addition, there was no multicollinearity between these four variables evaluated using the variance inflation factor (VIF) [63].

Table 1. List of independent variables.

Health behavior and clinical care variables
• Percentage of the population who smoke
• Percentage of the population who are obese
• Ratio of primary care physicians to population

Social and economic variables
• Percentage of adults aged 25–44 years with some post-secondary education
• Percentage of population aged 16 and older who are unemployed but seeking work
• Percentage of children that live in a single-parent household
• Number of membership associations per 10,000 population

Physical environmental variables
• Percentage of households with at least one of the following four housing problems: overcrowding, high housing costs, lack of kitchen or lack of plumbing facilities
• Mean county altitude in meters

Demographic variables
• Percentage of individuals aged 65 and over
• Percentage of African American
• Percentage of females
• Percentage of rural areas

Table 2. OLS regression analyses of all independent variables for suicide rates in the NCHS.

Model—With Mean County Altitude and Independent Variables—R-Squared = 0.485				
	Coefficient	S.E	t-Value	p-Value
Intercept	5.147	1.270	4.053	0.000 [a]
% smokers	0.312	0.019	16.556	0.000 [a]
% obesity	0.005	0.017	0.298	0.766
Primary care physician rate	0.002	0.002	1.381	0.167
% college education	0.006	0.006	1.101	0.271
% unemployment	0.008	0.029	0.292	0.771
% of single-parent households	0.006	0.007	0.825	0.410
Association rate	−0.053	0.009	−6.142	0.000 [a]
% severe housing problems	0.013	0.014	−0.913	0.361
% aged 65 and over	0.226	0.015	15.566	0.000 [a]
% African American	−0.036	0.005	−7.092	0.000 [a]
% female	−0.057	0.024	−2.318	0.021
% rural area	0.002	0.002	0.865	0.387
Mean county altitude	0.004	0.000	35.974	0.000 [a]

[a] Significant at $p < 0.001$.

2.4. Statistical Analyses

First, results from the data skewness analysis indicated that no data transformation was needed. Second, stepwise multiple regression was conducted to select significant predictor variables. Third, both OLS and GWR regression models were constructed to explore both global and local association between suicide rates and altitude, using the four predictor variables. An OLS model can be expressed as

$$\hat{y}_i = b_0 + \sum_k b_k x_{ik} + \varepsilon_i, \tag{1}$$

where $\hat{y}_i$ is the estimated value of the dependent variable for observation i, b_0 is the intercept, b_k is the coefficient for the predictor variable k, x_{ik} is the value of the kth predictor variable for observation i, and ε_i is the error term [64]. A GWR model, which allows examination of regional variation (spatial non-stationarity) in the relationship between predictor and dependent variables, can be written as

$$\hat{y}_i = b_0(u_i, v_i) + \sum_k b_k(u_i, v_i) x_{ik} + \varepsilon_i, \tag{2}$$

where $\hat{y}_i$ is the estimated value of the dependent variable for observation i, $b_0(u_i, v_i)$ denotes the intercept at the location i with coordinates (u_i, v_i); $b_k(u_i, v_i)$ represents the coefficient estimate for the predictor variable k at location i with coordinates of (u_i, v_i); x_{ik} denotes the observation on independent variable k at location i, and ε_i is the error term [64]. Instead of solving one single regression equation, GWR calibrates a separate regression equation for each observation, assuming that closer observations have greater influence on the estimation of the regression parameters than those further away ones [46]. The weight given to each observation is determined by a distance decay function.

Raster-based coefficient surfaces are then created using estimated coefficients at all the sample locations. A weight matrix is constructed for each raster cell, relating the location of the cell to the locations of all the other observations in the dataset. The weighting matrix is built based on a distance decay function, which is dependent on a chosen bandwidth. Two methods are commonly used to select the bandwidth: the cross-validation (CV) method and the Akaike Information Criteria (AIC) method [64]. Specifically, CV seeks the bandwidth that minimizes the CV score, expressed as

$$CV = \sum_{i=1}^{n} (y_i - \hat{y}_{\neq i})^2, \tag{3}$$

where n represents the number of observations. Note that observation i is excluded from the calibration when predicting the value at location i [63,64]. Alternatively, the AIC method finds the bandwidth that minimizes the AIC score, expressed as

$$AIC = 2n \log_e(\hat{\sigma}) + n \log_e(2\pi) + n \left\{ \frac{n + tr(S)}{n - 2 - tr(S)} \right\}, \tag{4}$$

where $tr(S)$ represents the trace of the hat matrix. The hat matrix describes the relationship between the fitted values and the observed values. The diagonal elements of the hat matrix specify the influence of each observed value on each fitted value for the same observation [65]. The AIC method has an advantage over the CV method in that it considers the degrees of freedom, which may vary between models centered on different observations [64]. In addition, bandwidth can either be fixed or adaptive. The fixed function uses an optimal bandwidth for the entire study area and all observations that fall within it are used in the subset regression. In contrast, the adaptive function allows bandwidth to vary based on the density of observation points [27,64,66]. Given the uneven distribution of counties in the data sets, the adaptive bandwidth with the AIC optimization method was applied in our analysis.

Moreover, since a GWR model provides separate parameter estimate measures, goodness-of-fit, and significance assessment for every observation in the dataset, regional variation (spatial non-stationarity) and significance of the association between the dependent variable and predictor variables, can be visualized and interpreted using maps made using the estimated values of the above measures [46].

3. Results

Table 3 shows the mean, standard deviations (SD), and range statistics of both dependent and independent variables. There was a total of 263,210 suicide deaths in the 3064 contiguous U.S. counties (or parishes) from 2008 to 2014. The seven-year smoothed age-adjusted average suicide rate per 100,000 population was 13.53 with a range from 5.17 to 70.14 and a SD of 3.53. The mean county altitude ranged between −0.37 m (above sea level) and 3473.11 m with a mean of 438.42 m and a SD of 509.71 m. The other four predictor variables all showed different degrees of variation across the states. On average, 18.39% of a county's population smoked with a SD of 3.71%. The African American population, on average, accounted for 12.39% of the county population with a SD of 14.33%. The proportion of a county's population aged 65 and above was, on average, 17.58% with a SD of 4.3%. The mean county association rate was 13.83 membership associations per 100,000 population with a SD of 6.78. The variations in these predictor variables indicate that suicide incidences occurred in regions with heterogeneous behavioral, socio-demographic, and environmental characteristics.

Table 3. Descriptive statistics for dependent and independent variables for NCHS.

	Mean	SD	Range
Suicide rates (per 100,000 population)	13.53	3.53	5.17–70.14
% smokers	18.39	3.71	6.90–41.20
% obesity	30.94	4.46	10.70–46.60
Primary care physician rate	55.74	34.75	0–469.23
% college education	56.27	11.65	2.70–100
% unemployment	6.26	2.30	1.20–23.70
% single-parent households	32.47	10.32	0–100
Association rate	13.83	6.78	0–81.30
% severe housing problems	14.47	4.86	2.18–71.26
% aged 65 and over	17.58	4.36	4.10–52.90
% African American	12.39	14.33	0–84.90
Association rate	13.83	6.78	0–81.30
% female	49.91	2.27	30.10–56.80
% rural area	58.82	31.50	0–100
Mean county altitude (m)	438.42	509.71	−0.37–3473.11

Figure 1 shows the suicide rates (smoothed and age-adjusted) in the contiguous U.S. counties from 2008 to 2014 using natural break classification [20]. Overall, suicide rates were much higher in the western counties compared to those in the other regions. In addition, the global Moran's *I* test (Moran's $I = 0.447$; $z = 87.94$; $p < 0.001$) suggested a spatially dependent county-level suicide pattern during our study period.

Table 4 summarizes the results from the OLS regression model. A R^2 value of 48.3% indicates that the five predictor variables explained almost half of the total variance in the county-level suicide rates. The VIF was less than two for all predictor variables, indicating no multicollinearity in the data. The suicide rates were significantly ($p < 0.001$) related to all five of the predictor variables. A positive relationship was found with the altitude, percentage of smoking population, and percentage of population aged 65 and above, while a negative relationship was discovered with the percentage of African American population and the association rate. Moreover, every 100 m increase in mean county altitude led to an increase in suicide rates of 0.4/100,000 ($B = 0.004$; $p = 0.000$), after adjusting the other four predictor variables. In addition, the model residuals were spatially autocorrelated (Moran's $I = 0.323$; $z = 63.526$; $p < 0.001$), meaning that the county-level suicide rates were spatially dependent across the 48 U.S. states. The residuals also indicated that suicide rates were underestimated primarily in the western states including Utah, Wyoming, Montana, South Dakota, and North Dakota (Figure 3).

Table 4. Partial results from the OLS regression model

Model—With Mean County Altitude and Potential Covariates—R-Squared = 0.483, AIC Value: 14,432.13				
	Coefficient	S.E	t-Value	p-Value
Intercept	2.730	0.336	8.134	0.000 [a]
Mean county altitude	0.004	0.000	40.691	0.000 [a]
% Smokers	0.334	0.013	25.196	0.000 [a]
% aged 65 and over	0.219	0.012	18.828	0.000 [a]
% African American	−0.036	0.004	−10.093	0.000 [a]
Association rate	−0.0045	0.007	−6.028	0.000 [a]

[a] Significant at $p < 0.001$.

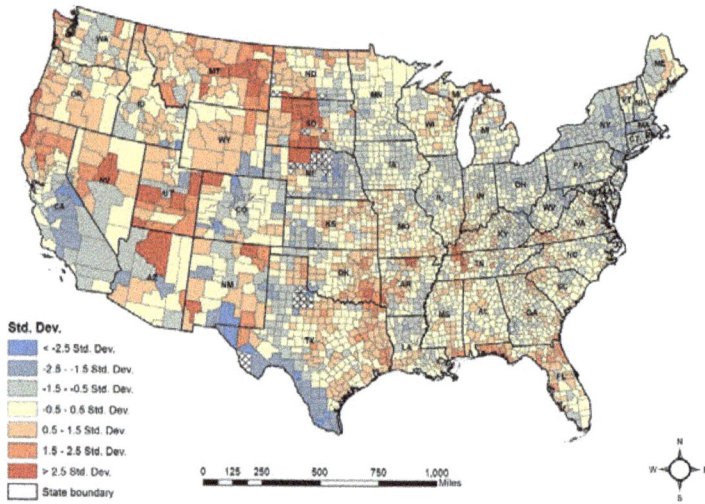

Figure 3. The residuals of the OLS model.

Table 5 presents the model fit measure along with the ranges of local coefficient estimates for the predictor variables from the GWR model. Compared with the results from the OLS model, the GWR model fit the data better with a higher R^2 value (average: 0.62; range: 0.21–0.64, Figure 4) and a lower AIC value (13,593.68). In addition, the spatial autocorrelation in the GWR model was significantly reduced, though was not totally eliminated, as indicated by the Moran's I statistic (Moran's I = 0.171; z = 33.656; $p < 0.001$).

Table 5. Partial results from the GWR model.

Model—With Mean County Altitude and Four Predictor Variables—R-Squared: 0.620, AIC Value: 13,593.68					
	Coefficient Range		Percentage of Counties by 95% of t Statistic		
	Min.	Max.	$t \leq -1.96$	$-1.96 < t < 1.96$	$t \geq 1.96$
Intercept	−5.68	10.41	6.62	16.75	76.63
Mean county altitude	−0.01	0.01	4.56	30.88	64.56
% Smokers	0.08	0.74	0.04	0	99.96
% 65 and over	0.11	0.31	0	0	100.00
% African American	−0.50	0.07	61.95	36.75	1.30
Association rate	−0.13	0.13	37.74	57.27	4.99

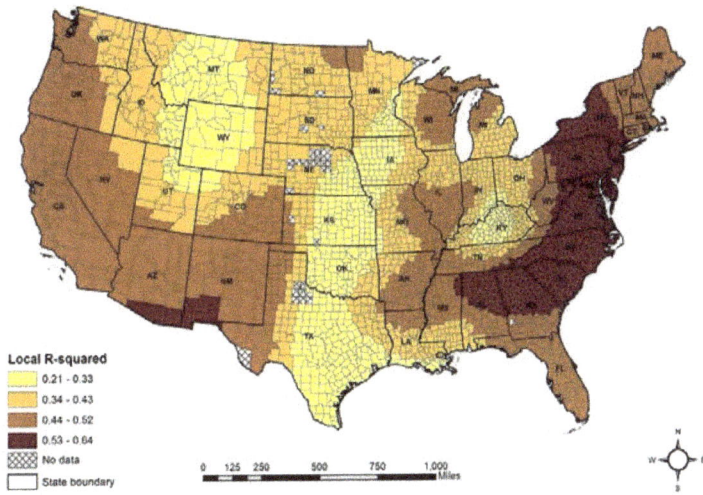

Figure 4. Local R^2 values from the GWR model.

Figure 5 shows the spatial patterns of the estimated coefficients of the five predictor variables. For smoking, stronger positive relationships were found in the northern plains, the Midwest, and northeast, particularly in portions of North Dakota, South Dakota, and Nebraska (Figure 5a). For percentage of the population aged 65 and above, stronger positive relationships could be seen in parts of the Midwest (North Dakota, South Dakota, and Nebraska), the southwest (Kansas, Oklahoma, New Mexico, and Texas), the northcentral (Michigan and eastern parts of Wisconsin), and portions of the central (Missouri, Illinois, Kentucky, and Tennessee) states (Figure 5b).

For mean county altitude, stronger positive relationships can be seen in portions of the northern and northern plains regions (Minnesota, North Dakota, and South Dakota), southeastern regions (West Virginia, Virginia, and North Carolina), and Louisiana. Stronger negative relationships were found in portions of Indiana and Illinois (Figure 5c). For the percentage of the African American population, stronger negative relationships appeared primarily in western states and positive relationships were discovered in the southwest (Texas, Kansas, and Nebraska) and the northeast (Kentucky, Ohio, Virginia, and West Virginia, and Pennsylvania) states (Figure 5d). For the association rate, stronger negative relationships were found in portions of the southwest, the north and the central north states, and positive relationships were seen in south Texas and all eastern states (Figure 5e).

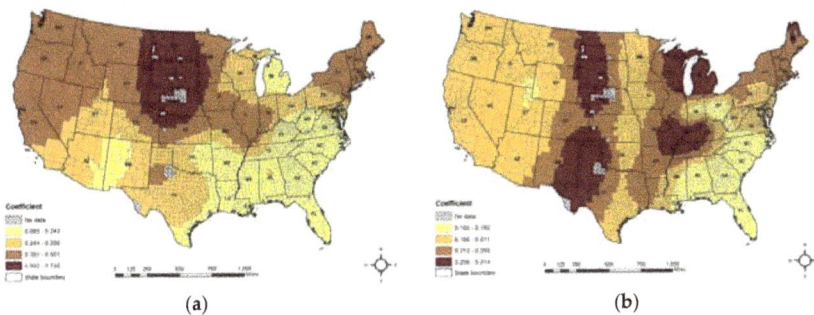

(a)

(b)

Figure 5. *Cont.*

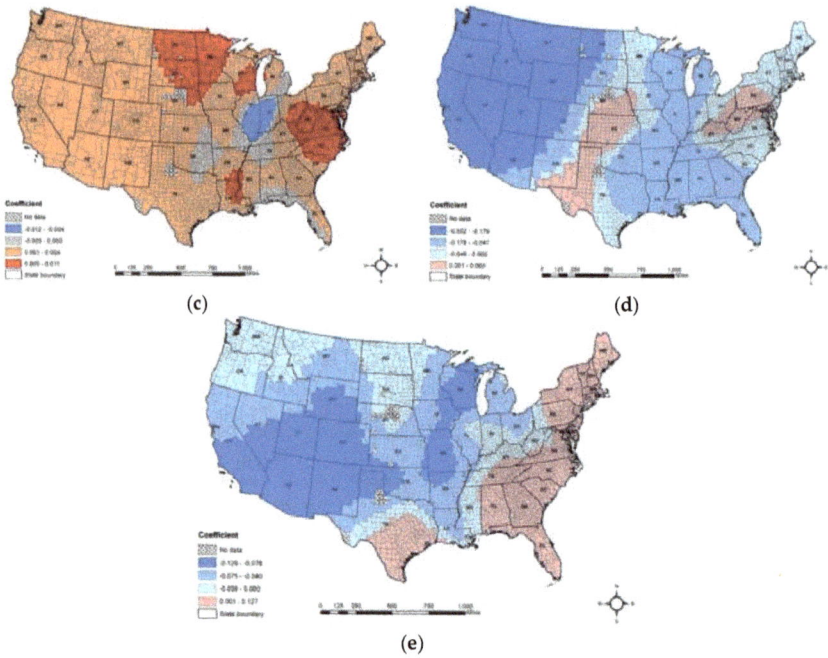

(c)

(d)

(e)

Figure 5. Coefficient estimates of the GWR model: (**a**) percentage of the local population that smoke; (**b**) percentage of the population aged 65 and above; (**c**) mean county altitude; (**d**) percentage of the African American population; and (**e**) the association rate.

Level of significance of the coefficients also varied across the 48 U.S. states. While consistent positive relationships were seen with smoking and elderly population, both negative and positive relationships were found with mean county altitude, percentage of the African American population, and the association rate. Moreover, mean county altitude had more positive relationships than the negative ones (64.56 percent vs. 4.56 percent) (Figure 6a). A negative relationship was found in 61.95 percent of the counties with of the percentage of the African American population (Figure 6b) and 37.74 percent of the counties with the association rate (Figure 6c). In addition, a positive relationship was found in 1.30 percent of the counties with the percentage of African American population (Figure 6b) and 4.99 percent of the counties with the association rate (Figure 6c).

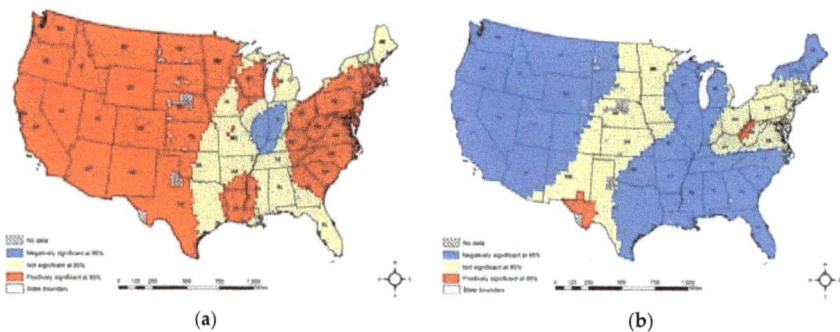

(a)

(b)

Figure 6. *Cont.*

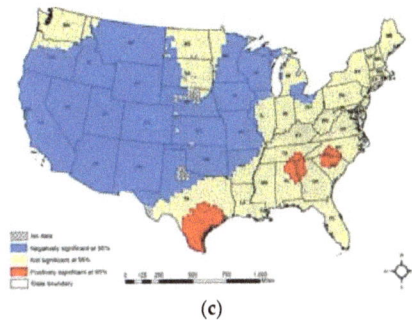

(c)

Figure 6. Significance map of the GWR model; (**a**) mean county altitude; (**b**) percentage of the African American population; and (**c**) the association rate.

4. Discussion

Our study expanded the existing literature on the association between suicide and altitude. A more recent U.S. suicide data set (2008–2014) was analyzed using both global and localized regression (OLS and GWR) models, and the mean county altitude was estimated with the best possible spatial resolution to our knowledge. Results from our GWR model showed that there existed a significant positive relationship between the county-level suicide rates and the county mean altitude, in close to two-thirds of the contiguous U.S. counties (particularly in the western and northeastern regions), after controlling for four predictor variables (i.e., percentage of the smoking population, percentage of population aged 65 and over, percentage of African American population African American, and the association rate). Our findings were generally consistent with the findings in the previous U.S. studies. However, the magnitude of the effect of altitude of our estimation was smaller than those in the literature [17,26,38]. Moreover, both unemployment and rurality were found to be insignificant in our models (Table 2) but were significant in most of the past studies [12,26,67,68].

In our study, we included all counties in our models so that continuous surfaces could be generated using the results from the GWR model to visualize the spatial variation of the relationship between the suicide rates and altitude across the contiguous U.S. Some scholars, concerned about the impacts of the floor effect [17], included only counties above a certain threshold altitude in their analyses. For instance, Huber et al. (2014) excluded counties with an altitude lower than 308.4 m (1000 ft.) from their models. As the floor effect may impact our modeling results, readers should interpret our modeling results with caution [53].

Our GWR model also identified areas where relationships between suicide rates and predictor variables were contradictory to most of the other counties. For instance, a pocket of counties in Illinois and Indiana showed a significant negative relationship between suicide and altitude. Some counties in northwest Texas and Ohio and West Virginia showed a significant positive relationship between suicide and the percentage of the American-African population. Significant positive relationships between suicide and the association rate were found in some areas in southeast Texas, north Alabama, south of North Carolina, and north of South Carolina. Several explanations are available to justify such relationships contradictory to mainstream theories as well as and the dominating patterns. Effects in these areas might be counteracted by local mechanisms that were not captured by the current model. For instance, a negative relationship between suicide and altitude might be related to the floor effect of the altitude in the area. One major advantage of using the GWR model is to identify regions with "abnormal" relationships so that more thorough research may be conducted in these areas.

Several integration mechanisms have been developed to explain the association between suicide rates and altitude. First, suicide is closely related to social relationships, social integration, and social regulation [28]. Thus, suicide occurs more frequently among those with weaker social connections and inadequate social support. A large portion of the population in the western U.S. are at a higher suicide risk because they reside at higher altitude regions that are much less socially connected. Second, it is well recognized that firearm ownership has been closely related to firearm suicide rates in the U.S. and thus, higher firearm ownership in the western U.S. may be a significant contributing factor for the higher suicide rates [30–32]. Third, from a biophysiological perspective, high altitude hypoxia may worsen mood, especially to those emotionally unstable, which may result in an increased suicide risk [58]. However, the relationship between altitude, hypoxia, mood and suicide needs further investigation. Some researchers argue that there lacks clinically proven evidence to substantiate a pathway connecting living in high altitude regions, hypoxia, worsening mood, depression, and suicide [69,70]. Previous studies also found that the relationship between high altitude and depression may be compensated by other factors. For instance, Ishikawa et al. (2013) found that the prevalence of depression was low in elderly highlanders living in the Himalayas and the Andes. The low prevalence was attributed to a deep devotion to religion and tight interpersonal networks [42]. In short, despite recent progress, robust pathways between altitude and suicide have yet to be established.

There are some limitations in this study. First, the reliability of the suicide data is questionable due to the well-known systematic under-reporting of suicide incidence [27,71]. Second, certain aspects of the GWR method, particularly regarding to the multicollinearity issue and approaches to calculating goodness of fit statistics, are still open to debate. For instance, some researchers contended that GWR might induce localized collinearity due to the inability of GWR to consistently differentiate the spatially stationary and non-stationary generating processes [72,73]. Fotheringham and Oshan [74], on the other hand, demonstrated that with controlled simulation GWR is sufficiently robust to withstand the multicollinearity effects [74]. For these reasons, GWR should be treated as more an exploratory data analysis tool to investigate the spatial non-stationarity relationship. Third, like most of the previous analyses, our study is threatened by ecological fallacy. Our data were aggregated at the county level, and individual-level risk factors were masked in our models. For example, residential history of individuals were not available so the altitude level an individual was residing at the time of the suicide was unknown, thus, the link between suicide and altitude is not robust. Lastly, other significant predictor variables including firearm ownership, religious affiliation, and accessibility of mental health services were not adjusted by our models because county-level data were not available when this study was conducted.

Furthermore, existing studies including ours have focused primarily on the spatial variations of suicide rates, but the temporal dimension of a spatial problem is at least equally important [75–77]. Geographical and temporal weighted regression (GTWR), an extension of GWR, has already been proposed to model the spatiotemporal pattern of local nonstationary processes [77,78]. In addition, past studies have suggested that specific meteorological conditions, such as atmospheric pressure, temperature, wind velocities, and hours of sunlight, might also be associated with suicide risk. However, the possible interaction between these factors and altitude was not considered in our analyses [79,80]. Finally, most of the previous studies pooled suicide data over several years to mitigate the small area problem. The assumption of this treatment of the raw data is that suicide is independent of time. However, if this assumption fails to hold, then the estimation of the suicide risk may be biased and spatial models for repeated time periods [81,82] and/or space-time models [83–85] are possible solutions for this problem.

5. Conclusions

We in this study examined the spatially varying relationship between county-level suicide rates and altitude using the most recent suicide data that are available in the U.S. Results from our regression models confirmed an overall positive relationship between county-level smoothed age-adjusted suicide rates and mean county altitude, after controlling for four predictor variables including smoking, age of population, percentage of African American population, and the association rate. The relationships varied considerably across the contiguous U.S. Counties in the western and northeastern regions appeared to mostly align well with the hypothesized positive relationship between altitude and suicide risk, although a pocket of counties in Illinois and Indiana showed a reversed relationship. Future research may further investigate areas with a negative relationship as well as testing additional predictor variables, such as religious affiliation and accessibility of mental health facilities and services. Future research may also simultaneously model the varying relationship in space and over time. We hope that a better understanding of the association between suicide and altitude may help identify high-risk regions for further research and/or for formulating targeted prevention and intervention strategies.

Acknowledgments: The authors would like to thank the anonymous reviewers for their constructive comments and suggestions to improve the paper. We are grateful for research assistance provided by Lingling Chen. The contents of this publication are solely the responsibility of the authors.

Author Contributions: Hoehun Ha performed the statistical analyses, and wrote the manuscript. Wei Tu provided advice on data analysis and also contributed to the writing of the manuscript.

Conflicts of Interest: The authors declare no conflict of interest.

References

1. Mathers, C.D.; Loncar, D. Projections of global mortality and burden of disease from 2002 to 2030. *PLoS Med.* **2006**, *3*, 2011–2030. [CrossRef] [PubMed]
2. World Health Organization. *World Report on Violence and Health*; World Health Organization: Geneva, Switzerland, 2002.
3. World Health Organization (WHO). Suicide Data. 2017. Available online: http://www.who.int/mental_health/prevention/suicide/suicideprevent/en/ (accessed on 7 September 2017).
4. Florence, C.; Haegerich, T.; Simon, T.; Zhou, C.; Luo, F. Estimated lifetime medical and work-loss costs of emergency department–treated nonfatal injuries—United States, 2013. *Morb. Mortal. Wkly. Rep.* **2015**, *64*, 1078–1082. [CrossRef] [PubMed]
5. Shepard, D.S.; Gurewich, D.; Lwin, A.K.; Reed, G.A., Jr.; Silverman, M.M. Suicide and suicidal attempts in the United States: Costs and policy implications. *Suicide Lift Threat Behav.* **2016**, *46*, 352–362. [CrossRef] [PubMed]
6. The National Institute of Mental Health (NIMH). Suicide. 2017. Available online: https://www.nimh.nih.gov/health/statistics/suicide/index.shtml (accessed on 7 September 2017).
7. Daly, M.C.; Wilson, D.J.; Johnson, N.J. Relative status and well-being: Evidence from U.S. Suicide deaths. *Rev. Econ. Stat.* **2007**, *95*, 1480–1500. [CrossRef]
8. Dumais, A.; Lesage, A.D.; Alda, M.; Rouleau, G.; Dumont, M.; Chawky, N.; Roy, M.; Mann, J.J.; Benkelfat, C.; Turecki, G. Risk factors for suicide completion in major depression: A case-control study of impulsive and aggressive behaviors in men. *Am. J. Psychiatry* **2005**, *162*, 2116–2124. [CrossRef] [PubMed]
9. Hughes, J.R. Smoking and suicide: A brief overview. *Drug Alcohol Depend.* **2008**, *98*, 169–178. [CrossRef] [PubMed]
10. Poorolajal, J.; Darvishi, N. Smoking and Suicide: A Meta-Analysis. *PLoS ONE* **2016**, *11*, e0156348. [CrossRef] [PubMed]
11. Lawrence, R.E.; Oquendo, M.A.; Stanley, B. Religion and Suicide Risk: A Systematic Review. *Arch. Suicide Res.* **2016**, *20*, 1–21. [CrossRef] [PubMed]
12. Singh, G.K.; Siahpush, M. Increasing rural–urban gradients in US suicide mortality, 1970–1997. *Am. J. Public Health* **2002**, *92*, 1161–1167. [CrossRef] [PubMed]
13. Stack, S. Suicide: A 15-year review of the sociological literature part I: Cultural and economic factors. *Suicide Life Threat Behav.* **2000**, *30*, 145–162. [PubMed]
14. Deisenhammer, E.A.; Kemmler, G.; Parson, P. Association of meteorological factors with suicide. *Acta Psychiatr. Scand.* **2003**, *108*, 455–459. [CrossRef] [PubMed]
15. Young, S. Elevated incidence of suicide in people living at altitude, smokers and patients with chronic obstructive pulmonary disease and asthma: Possible role of hypoxia causing decreased serotonin synthesis. *J. Psychiatry Neurosci.* **2013**, *38*, 423–426. [CrossRef] [PubMed]

16. Lin, G.Z.; Li, L.; Song, Y.F.; Zhou, Y.X.; Shen, S.Q.; Ou, C.Q. The impact of ambient air pollution on suicide mortality: A case-crossover study in Guangzhou, China. *Environ. Health* **2016**, *15*, 90. [CrossRef] [PubMed]
17. Brenner, B.; Cheng, D.; Clark, S.; Camargo, C.A., Jr. Positive association between altitude and suicide in 2584 US counties. *High Alt. Med. Biol.* **2011**, *12*, 31–35. [CrossRef] [PubMed]
18. Nock, M.K.; Park, J.M.; Finn, C.T.; Deliberto, T.L.; Dour, H.J.; Banaji, M.R. Measuring the suicidal mind: Implicit cognition predicts suicidal behavior. *Psychol. Sci.* **2010**, *21*, 511–517. [CrossRef] [PubMed]
19. Strandheim, A.; Bjerkeset, O.; Gunnell, D.; Bjørnelv, S.; Holmen, T.L.; Bentzen, N. Risk factors for suicidal thoughts in adolescence-a prospective cohort study: The Young-HUNT study. *BMJ Open* **2014**, *4*, e005867. [CrossRef] [PubMed]
20. Abdeen, Z.; Brunstein-Klomek, A.; Nakash, O.; Shibli, N.; Nagar, M.; Agha, H.; Hallaq, S.; Kanat-Maymon, Y.; Juerges, H.; Levav, I.; et al. The Association between Political Violence and the Connection between Bullying and Suicidality among Palestinian Youth. *Suicide Life Threat Behav.* **2017**. [CrossRef] [PubMed]
21. Harper, S.; Charters, T.J.; Strumpf, E.C.; Galea, S.; Nandi, A. Economic downturns and suicide mortality in the USA, 1980–2010: Observational study. *Int. J. Epidemiol.* **2015**, *44*, 956–966. [CrossRef] [PubMed]
22. Norstrom, T.; Gronqvist, H. The Great Recession, unemployment and suicide. *J. Epidemiol. Community Health* **2015**, *69*, 110–116. [CrossRef] [PubMed]
23. U.S. Centers for Disease Control Prevention (U.S. CDC). Fatal Injury Reports, National, Regional and State, 1981–2015. Atlanta, CDC, Webbased Injury Statistics Query and Reporting System (WISQARS). 2017. Available online: https://webappa.cdc.gov/sasweb/ncipc/mortrate.html (accessed on 7 September 2017).
24. Lester, D. Variation in suicide and homicide rates by latitude and longitude in the United States, Canada, and Australia. *Am. J. Psychiatry* **1985**, *142*, 523–524. [PubMed]
25. Chang, S.S.; Sterne, J.A.; Wheeler, B.W.; Lu, T.H.; Lin, J.J.; Gunnell, D. Geography of suicide in Taiwan: Spatial patterning and socioeconomic correlates. *Health Place* **2011**, *17*, 641–650. [CrossRef] [PubMed]
26. Kim, N.; Mickelson, J.B.; Brenner, B.E.; Haws, C.A.; Yurgelun-Todd, D.A.; Renshaw, P.F. Altitude, gun ownership, rural areas, and suicide. *Am. J. Psychiatry* **2011**, *168*, 49–54. [CrossRef] [PubMed]
27. Trgovac, A.B.; Kedron, P.J.; Bagchi-Sen, S. Geographic variation in male suicide rates in the United States. *Appl. Geogr.* **2015**, *62*, 201–209. [CrossRef]
28. Bearman, P.S. The social structure of suicide. *Sociol. Forum* **1991**, *6*, 501–524. [CrossRef]
29. Durkheim, E.; Simpson, G. *Suicide: A Study in Sociology*; Routledge & Kegan Paul: London, UK, 1952.
30. Anestis, M.D.; Houtsma, C. The Association between Gun Ownership and Statewide Overall Suicide Rates. *Suicide Life Threat Behav.* **2017**. [CrossRef] [PubMed]
31. Miller, M.; Hemenway, D. Guns and suicide in the United States. *N. Engl. J. Med.* **2008**, *359*, 989–991. [CrossRef] [PubMed]
32. Siegel, M.; Rothman, E.F. Firearm Ownership and Suicide Rates among US Men and Women, 1981–2013. *Am. J. Public Health* **2016**, *106*, 1316–1322. [CrossRef] [PubMed]
33. Guger, C.; Krausert, S.; Domej, W.; Edlinger, G.; Tannheimer, M. EEG, ECG and oxygen concentration changes from sea level to a simulated altitude of 4000 m and back to sea level. *Neurosci. Lett.* **2008**, *442*, 123–127. [CrossRef] [PubMed]
34. Katz, I.R. Is there a hypoxic affective syndrome? *Psychosomatics* **1982**, *23*, 846–853. [CrossRef]
35. Shukitt-Hale, B.; Banderet, L.E.; Lieberman, H.R. Elevation-dependent symptom, mood, and performance changes produced by exposure to hypobaric hypoxia. *Int. J. Aviat. Psychol.* **1998**, *8*, 319–334. [CrossRef] [PubMed]
36. Kious, B.M.; Kondo, D.G.; Renshaw, P.F. Living High and Feeling Low: Altitude, Suicide, and Depression. *Harv. Rev. Psychiatry* **2018**, *26*, 43–56. [CrossRef] [PubMed]
37. Reno, E.; Brown, T.L.; Betz, M.E.; Allen, M.H.; Hoffecker, L.; Reitinger, J.; Roach, R.; Honigman, B. Suicide and high altitude: An integrative review. *High Alt. Med. Biol.* **2017**. [CrossRef] [PubMed]
38. Kim, J.; Choi, N.; Lee, Y.J.; An, H.; Kim, N.; Yoon, H.K.; Lee, H.-J. High altitude remains associated with elevated suicide rates after adjusting for socioeconomic status: A study from South Korea. *Psychiatry Investig.* **2014**, *11*, 492–494. [CrossRef] [PubMed]
39. Ortiz-Prado, E.; Simbaña, K.; Gómez, L.; Henriquez-Trujillo, A.R.; Cornejo-Leon, F.; Vasconez, E.; Castillo, D.; Viscor, G. The disease burden of suicide in Ecuador, a 15 years' geodemographic cross-sectional study (2001–2015). *BMC Psychiatry* **2017**, *17*, 342. [CrossRef] [PubMed]
40. Selek, S. Altitude, immigration and suicide rates: A study from Turkey. *Psychiatry Investig.* **2013**, *10*, 89–91. [CrossRef] [PubMed]

41. Oka, M.; Kubota, T.; Tsubaki, H.; Yamauchi, K. Analysis of impact of geographic characteristics on suicide rate and visualization of result with Geographic Information System. *Psychiatry Clin. Neurosci.* **2015**, *69*, 375–382. [CrossRef] [PubMed]
42. Ishikawa, M.; Yamanaka, G.; Nakajima, S.; Suwa, K.; Matsuda, A.; Nakaoka, T.; Okumiya, K.; Matsubayashi, K.; Otsuka, K. Association between high altitude and depression in the Himalayas and the Andes. *Nihon Ronen Igakkai Zasshi* **2013**, *50*, 330–334. [CrossRef] [PubMed]
43. Haws, C.A.; Gray, D.D.; Yurgelun-Todd, D.A.; Moskos, M.; Meyer, L.J.; Renshaw, P.F. The possible effect of altitude on regional variation in suicide rates. *Med. Hypotheses* **2009**, *73*, 587–590. [CrossRef] [PubMed]
44. Goovaerts, P. Geostatistical analysis of health data: State-of-the-art and perspectives. *Geostat. Environ. Appl.* **2008**, *15*, 3–22.
45. Leung, Y. Discovery of Spatial Relationships in Spatial Data. In *Knowledge Discovery in Spatial Data*; Springer: Berlin, Germany, 2010; pp. 223–276.
46. Fotheringham, A.S.; Brunsdon, C.; Charlton, M. *Geographically Weighted Regression: The Analysis of Spatially Varying Relationships*; Wiley: Hoboken, NJ, USA, 2002.
47. Chan, T.C.; Chiang, P.H.; Su, M.D.; Wang, H.W.; Liu, M.S. Geographic disparity in chronic obstructive pulmonary disease (COPD) mortality rates among the Taiwan population. *PLoS ONE* **2014**, *9*, e98170. [CrossRef] [PubMed]
48. Lin, C.H.; Wen, T.H. Using geographically weighted regression (GWR) to explore spatial varying relationships of immature mosquitoes and human densities with the incidence of dengue. *Int. J. Environ. Res. Public Health.* **2011**, *8*, 2798–2815. [CrossRef] [PubMed]
49. Shoff, C.; Yang, T.C. Spatially varying predictors of teenage birth rates among counties in the United States. *Demogr. Res.* **2012**, *27*, 377–418. [CrossRef] [PubMed]
50. Tu, J.; Tu, W.; Tedders, S.H. Spatial variations in the associations of term birth weight with ambient air pollution in Georgia, USA. *Environ. Int.* **2016**, *92*, 146–156. [CrossRef] [PubMed]
51. Helbich, M.; Leitner, M.; Kapusta, N.D. Geospatial examination of lithium in drinking water and suicide mortality. *Int. J. Helath Geogr.* **2012**, *11*, 19. [CrossRef] [PubMed]
52. Cheng, D.C.; Mendenhall, T.I.; Barry, B. Suicide rates strongly correlate with altitude. *Acad. Emerg. Med.* **2005**, *12* (Suppl. 1), 141. [CrossRef]
53. Huber, R.; Coon, H.; Kim, N.; Renshaw, P.F.; Kondo, D.G. Altitude is a Risk Factor for Completed Suicide in Bipolar Disorder. *Med. Hypotheses* **2014**, *82*, 377–381. [CrossRef] [PubMed]
54. National Center for Health Statistics (NCHS). *WISQARS Fatal Injury Mortality Reports, 1999–2007*; National Center for Inquiry Prevention and Control: Atlanta, GA, USA, 2011.
55. U.S. Geological Survey (USGS). 100-Meter Resolution Elevation of the Conterminous United States. 2012. Available online: https://catalog.data.gov/dataset/100-meter-resolution-color-sliced-elevation-of-the-conterminous-united-states-direct-download (accessed on 10 September 2017).
56. ESRI. ArcGIS 10.3.1. 2015. Available online: http://desktop.arcgis.com/en/arcmap/10.3/main/get-started/whats-new-in-arcgis-1031.htm (accessed on 7 October 2017).
57. The U.S. Census Bureau. Geography: Cartographic Boundary Shapefiles—Counties. 2017. Available online: https://www.census.gov/geo/maps-data/data/cbf/cbf_counties.html (accessed on 8 September 2017).
58. Ha, H. Geographic variation in Mentally Unhealthy Days: Air pollution and altitude perspectives. *High Alt. Med. Biol.* **2017**, *18*, 258–266. [CrossRef] [PubMed]
59. The U.S. Census Bureau. Population and Housing Unit Estimates. 2017. Available online: https://www.census.gov/programs-surveys/popest.html (accessed on 8 September 2017).
60. U.S. Centers for Disease Control Prevention (U.S. CDC). SMART: BRFSS City and County Data and Documentation. 2017. Available online: https://www.cdc.gov/brfss/smart/Smart_data.htm (accessed on 7 December 2016).
61. Voss, J.D.; Masuoka, P.; Webber, B.J.; Scher, A.I.; Atkinson, R.L. Association of elevation, urbanization and ambient temperature with obesity prevalence in the United States. *Int. J. Obes.* **2013**, *37*, 1407–1412. [CrossRef] [PubMed]
62. The U.S. Census Bureau. The American Community Survey. 2017. Available online: https://www.census.gov/programs-surveys/acs/ (accessed on 8 September 2017).
63. Rogerson, P.A. *Statistical Methods for Geography*, 2nd ed.; Sage: London, UK, 2006.
64. Mennis, J. Mapping the results of geographically weighted regression. *Cartogr. J.* **2006**, *43*, 171–179. [CrossRef]

65. Hoaglin, D.C.; Welsch, R.E. The hat matrix in regression and ANOVA. *Am. Stat.* **1978**, *32*, 17–22.
66. Nakaya, T.; Charlton, M.; Fotheringham, A.S.; Brunsdon, C. Geographically Weighted Regression Version 4. User's Manual and Installation Guide. 2009. Available online: http://gwr.maynoothuniversity.ie/wp-content/uploads/2013/04/GWR4_Manual.pdf (accessed on 8 September 2017).
67. Fontanella, C.A.; Hiance-Steelesmith, D.L.; Phillips, G.S.; Bridge, J.A.; Lester, N.; Sweeney, H.A.; Campo, J.V. Widening rural-urban disparities in youth suicides, United States, 1996–2010. *JAMA Pediatr.* **2015**, *169*, 466–473. [CrossRef] [PubMed]
68. Hirsch, J.K. A review of the literature on rural suicide: Risk and protective factors, incidence, and prevention. *Crisis* **2006**, *27*, 189–199. [CrossRef] [PubMed]
69. Ha, H. The Conversation. The Curious Relationship between Altitude and Suicide. 2017. Available online: https://theconversation.com/the-curious-relationship-between-altitude-and-suicide-85716 (accessed on 5 November 2017).
70. Betz, M.E.; Valley, M.A.; Lowenstein, S.R.; Hedegaard, H.; Thomas, D.; Stallones, L.; Honigman, B. Elevated suicide rates at high altitude: Sociodemographic and health issues may be to blame. *Suicide Life Threat Behav.* **2011**, *41*, 562–573. [CrossRef] [PubMed]
71. Tøllefsen, I.M.; Hem, E.; Ekeberg, Ø. The reliability of suicide statistics: A systematic review. *BMC Psychiatry* **2012**, *12*, 9. [CrossRef] [PubMed]
72. Páez, A.; Farber, S.; Wheeler, D. A simulation-based study of geographically weighted regression as a method for investigating spatially varying relationships. *Environ. Plan. A* **2011**, *43*, 2992–3010. [CrossRef]
73. Wheeler, D.; Tiefelsdorf, M. Multicollinearity and correlation among local regression coefficients in geographically weighted regression. *J. Geogr. Syst.* **2005**, *7*, 161–187. [CrossRef]
74. Fotheringham, A.S.; Oshan, T.M. Geographically weighted regression and multicollinearity: Dispelling the myth. *J. Geogr. Syst.* **2016**, *18*, 303–329. [CrossRef]
75. MacNab, Y.; Dean, C.B. Spatio-temporal modelling of rates for the construction of disease maps. *Stat. Med.* **2002**, *21*, 347–358. [CrossRef] [PubMed]
76. Meliker, J.R.; Sloan, C.D. Spatio-temporal epidemiology: Principles and opportunities. *Spat. Spatio-temp. Epudemiol.* **2011**, *2*, 1–9. [CrossRef] [PubMed]
77. Fotheringham, A.S.; Crespo, R.; Yao, J. Geographical and Temporal Weighted Regression (GTWR). *Geogr. Anal.* **2015**, *47*, 431–452. [CrossRef]
78. Huang, B.; Wu, B.; Barry, M. Geographically and temporally weighted regression for modeling spatio-temporal variation in house prices. *Int. J. Geogr. Inf. Sci.* **2010**, *24*, 383–401. [CrossRef]
79. Jessen, G.; Steffensen, P.; Jensen, B. Seasons and meteorological factors in suicidal behaviour. *Arch. Suicide Res.* **1998**, *4*, 263–280. [CrossRef]
80. Deisenhammer, E.A. Weather and suicide: The present state of knowledge on the association of meteorological factors with suicidal behaviour. *Acta Psychiatr. Scand.* **2003**, *108*, 402–409. [CrossRef] [PubMed]
81. Gunnell, D.; Wheeler, B.; Chang, S.S.; Thomas, B.; Sterne, J.A.; Dorling, D. Changes in the geography of suicide in young men: England and Wales 1981–2005. *J. Epidemiol. Community Health* **2012**, *66*, 536–543. [CrossRef] [PubMed]
82. Santana, P.; Costa, C.; Cardoso, G.; Loureiro, A.; Ferrão, J. Suicide in Portugal: Spatial determinants in a context of economic crisis. *Health Place* **2015**, *35*, 85–94. [CrossRef] [PubMed]
83. Abellan, J.J.; Richardson, S.; Best, N. Use of space–time models to investigate the stability of patterns of disease. *Environ. Health Perspect.* **2008**, *116*, 1111. [CrossRef] [PubMed]
84. Blangiardo, M.; Cameletti, M.; Baio, G.; Rue, H. Spatial and spatio-temporal models with R-INLA. *Spat. Spatio-temp. Epidemiol.* **2013**, *7*, 39–55. [CrossRef]
85. Helbich, M.; Plener, P.L.; Hartung, S.; Blüml, V. Spatiotemporal Suicide Risk in Germany: A Longitudinal Study 2007–11. *Sci. Rep.* **2017**, *7*, 7673. [CrossRef] [PubMed]

International Journal of
*Environmental Research
and Public Health*

MDPI

Article

The Relationship between Air Pollution and Depression in China: Is Neighbourhood Social Capital Protective?

Ruoyu Wang [1,2], Desheng Xue [1,2], Ye Liu [1,2,*] ⓘ, Penghua Liu [1,2] and Hongsheng Chen [3]

[1] School of Geography and Planning, Sun Yat-Sen University, Xingang Xi Road, Guangzhou 510275, China; wangry6@mail2.sysu.edu.cn (R.W.); eesxds@mail.sysu.edu.cn (D.X.); liuph3@mail2.sysu.edu.cn (P.L.)
[2] Guangdong Key Laboratory for Urbanization and Geo-simulation, Sun Yat-Sen University, Xingang Xi Road, Guangzhou 510275, China
[3] School of Architecture, Southeast University, Si-Pai-Lou Road No. 2, Nanjing 210096, China; hongsheng.chen2006@163.com
* Correspondence: liuye25@mail.sysu.edu.cn

Received: 30 April 2018; Accepted: 31 May 2018; Published: 2 June 2018

Abstract: There is increasing evidence from the developed world that air pollution is significantly related to residents' depressive symptoms; however, the existence of such a relationship in developing countries such as China is still unclear. Furthermore, although neighbourhood social capital is beneficial for health, whether it is a protective factor in the relationship between health and environment pollution remains unclear. Consequently, we examined the effects of cities' $PM_{2.5}$ concentrations on residents' depressive symptoms and the moderating effects of neighbourhood social capital, using data from the 2016 wave of China Labourforce Dynamics Survey and the real-time remote inquiry website of Airborne Fine Particulate Matter and Air Quality Index. Results showed that $PM_{2.5}$ concentrations and neighbourhood social capital may increase and decrease respondents' depressive symptoms, respectively. Notably, neighbourhood social capital decreased the negative effect of $PM_{2.5}$ concentrations on respondents' depressive symptoms. These analyses contributed to the understanding of the effect of air pollution on mental health in China and confirmed that neighbourhood social capital were protective factors in the relationship between health and environment hazards.

Keywords: $PM_{2.5}$ concentrations; depressive symptoms; neighbourhood social capital; moderate effect; China

1. Introduction

Depressive symptoms are a fundamental problem globally and are considered one of the most severe mental health problems [1]. As of 2010, depressive disorder was the 11th leading cause of disability-adjusted life years globally [2]. Depression is not only associated with physical illnesses like cardiovascular problems [3,4], decreased quality of life [5], and decreased work productivity [6,7], but also increases the mortality and suicide rate [3,4]. Data from the China Health and Retirement Longitudinal Study showed that the rate of depression among the elderly in China reached 31.2% in 2013 [8]. Therefore, depression has become a notable problem in China. Most previous research has indicated that depressive symptoms are significantly related to socioeconomic status and health-related behaviour [2–5]. For example, people with higher educational attainment and household income are less likely to suffer from depression than those who are less affluent and educated [2–5]. Similarly, people who drink alcohol or smoke frequently and engage in fewer physical activities are more likely to suffer from depression than those who refrain from drinking, smoking and regularly exercise [2–5].

Recent studies in developed countries have reported that higher air pollution may decrease residents' mental health, especially increasing the risk of experiencing depression [9–16]. Although the biological mechanisms for this association are not completely understood [12], possible pathological pathways are that: (1) air pollution may increase the risk of cardiovascular illness and thus increases the risk of depression [11,12,17]; and (2) air pollution influences mental health by affecting the nervous and digestive system [12,18]. In addition, another possible biological pathway is that air pollution may reduce the volume of sunlight which may be a stressor for residents' nervous system and contributes to the development of depression [19–21]. Besides biological effects, air pollution may also affect depression through other health-related behaviours. One possible pathway is that air pollution may pose a barrier to involvement with outdoor physical exercises, which is an effective way to cope with the risk of mental health problems [16,22–24]. Another explanation is associated with access to face-to-face social contact. Some studies have shown that air pollution discouraged face-to-face social contact among neighbours and therefore increased the risk for depression [25–27]. Although a growing body of literature has investigated the health effect of Particulate Matter 2.5 ($PM_{2.5}$) in Chinese cities, this body of research is largely related to the effects on physical health rather than those on mental health [9,10].

Social capital has been defined as a type of resource that provides people with convenience, and it has become essential for the maintenance of population health over the last two decades [28]. Thus, many studies have reported that social capital including social trust, social reciprocity, and social group membership can improve residents' health, as people living in neighbourhood with higher neighbourhood social capital could more easily acquire emotional or material support from others [1,28]. In recent years, social capital research has posited that neighbourhood social capital may also be protective for mental health, as it weakens the negative influences of neighbourhood social hazards [29–37]. For example, Feng et al. have noted that neighbourhood social trust can improve residents' mental health, since people in communities with high neighbourhood social trust can get more health information [31]. Lindström et al. have pointed out that neighbourhood social group membership and participation may benefit residents' mental health, as residents are more likely to get support from other neighbourhood group members [32]. Abbott et al. have found that neighbourhood social reciprocity may increase people's health by strengthening social ties and regulating residents' health related behaviours [33]. Further, neighbourhood social capital may be a buffer between social environmental hazards and residents' mental health [29,34–37]. For example, Niedzwiedz et al. have indicated that neighbourhood group membership may act as a buffer between the inequity of household income and older people's mental health [29]. Murayama et al. have also reported that the presence of both neighbourhood social trust and social reciprocity can weaken the negative effects of neighbourhood social environment on residents' depressive mood [37]. However, other researchers have recently argued that social capital may not only act as a protective factor for social environmental hazards but also for physical environmental hazards [38]. For instance, social capital may moderate the effect of industrial pollution exposure on residents' self-reported health because air pollution may contribute to feelings of stress in residents, and social capital can be a buffer its effects [38].

Although earnest efforts have been adopted to investigate the relationship between air pollution and depressive symptoms in developed countries, the relationship is still unclear in developing countries, such as China. Furthermore, although there is a consensus that neighbourhood social capital is beneficial for mental health, its protective benefits in the relationship between depressive symptoms and air pollution is still unclear. To bridge these gaps, this study investigated the relationship between $PM_{2.5}$ concentrations and depressive symptoms in China using data from the 2016 wave of the China Labourforce Dynamics Survey (CLDS 2016). We further examine whether neighbourhood social capital has a protective influence on the relationship between depressive symptoms and PM2.5 concentrations. This study contributes to the body of literature in two respects: first, it improves our understanding of how air pollution negatively influences people's mental health in China; and, second, it provides a deeper understanding of the protective function of neighbourhood social capital for mental health. The conceptual model is presented in Figure 1.

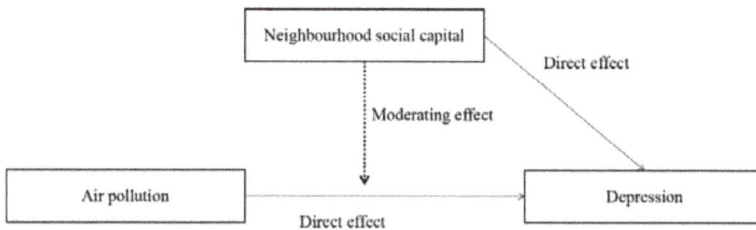

Figure 1. Conceptual model of the moderating relationship of neighbourhood social capital on the relationship between air pollution and depressive symptoms.

Based on the conceptual model and the review of existing literature, we propose the following hypotheses:

Hypothesis 1. *Residents who live in cities with higher levels of air pollution are more likely to have higher levels of depressive symptoms compared to residents in other cities.*

Hypothesis 2. *Residents who live in neighbourhood with higher neighbourhood social capital are likely to have lower levels of depressive symptoms than residents who live in neighbourhood with lower social capital.*

Hypothesis 3. *Neighbourhood social capital weakens the negative effect of air pollution on residents' depressive symptoms.*

2. Data and Methods

2.1. Data

The CLDS 2016 conducted by the Centre for Social Science Survey of Sun Yat-sen University [39] was the primary source of the data. Respondents from this survey were chosen by using a probability proportional to size sampling technique. First, 158 prefecture-level divisions from 29 provinces were randomly selected. Second, 401 neighbourhoods were randomly chosen from the prefecture-level divisions. In China, neighbourhoods refer to basic administrative divisions nested within prefectures (prefecture-level division means the second-level administrative divisions). Overall, 20,861 individuals nested within 401 neighbourhoods nested within 158 prefectures were included in the final dataset.

Average annual $PM_{2.5}$ concentrations of each prefecture were acquired from a real-time remote inquiry website—Airborne Fine Particulate Matter and Air Quality Index [40]—which provides a quantitative hourly index of air pollutants such as AQI, CO, NO_2, SO_2, O_3, PM_{10}, and $PM_{2.5}$. By calling the API of the website, the AQI observation data from January to December 2015 in 1613 monitoring stations were collected and cleaned. However, there was a lot of missing data and noise in the observation data. Therefore, a Kalman filter was used for the optimal estimate of $PM_{2.5}$ observation values [41]. A Kalman filter is a time-domain filtering algorithm, which can obtain the optimal estimation of the next moment based on the state of the system and the observation value at the next moment [42]. Another problem was that the recorded data had neither longitude nor latitude information, which may make it difficult to represent spatial variation. To map the data into a geographical space, geocoding was used to associate the observation data with spatial coordinates. Figure 2a shows the locations of 1613 monitoring stations in China in 373 cities. Further, the ordinary kriging interpolation method was used to generate a continuous raster surface (pixel size = 500 m), and zonal statistics was applied to extract the mean concentrations of PM2.5 in each prefecture. Figure 2b shows the spatial interpolation of the average $PM_{2.5}$ concentration in 2015, which was obtained by applying the Kriging method to the ground-based observations average in 2015.

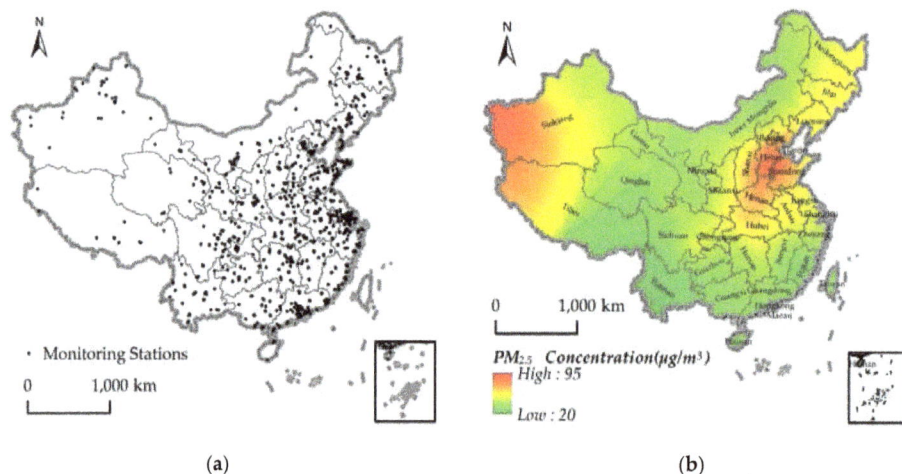

Figure 2. (a) Locations of monitoring stations; (b) PM2.5 measurement in January 2015.

2.2. Variables

The Center for Epidemiologic Studies Depression scale (CES-D) was used to calculate depressive symptoms, which has been widely used in previous studies [1]. The CES-D has a Cronbach's alpha value of 0.95. Independent variables included $PM_{2.5}$ concentrations and social capital indicators. Following existing studies ,the average annual $PM_{2.5}$ concentrations of each prefecture were used to measure the severity of air pollution in cities in 2015 [11,12].

Based on the indicators used in previous studies, three main social capital indicators (i.e social trust, social reciprocity, and social group membership) were included [28]. These three social capital indicators were measured based on variables included in the questionnaire (Part six: Social participation and support) that was collected in CLDS 2016. The question 'Would you say most people can be trusted?' was used to measure respondents' perceptions of social trust. The response '*Neighbours are extremely/very trustworthy*' was defined as high social trust. The question 'Would you say most of the time people try to be helpful?' was used to assess respondents' perceptions of social reciprocity. The response '*Neighbours always/often help each other*' was defined as high social reciprocity. Lastly, respondents were asked about membership in a various kind of voluntary groups. Following existing studies, the percentage of respondents being high trust and high reciprocity within each neighbourhood were calculated and defined as aggregated neighbourhood social trust and aggregated neighbourhood social reciprocity [22]. Furthermore, the average number of types of voluntary groups within each neighbourhood was calculated and defined as aggregated neighbourhood social group membership.

Lastly, we controlled for a series of individual-level and neighbourhood-level variables including: gender (dichotomous variable), age (continuous variable), marital status (categorical variables), educational attainment (categorical variables), employment status (dichotomous variable), *hukou* status (dichotomous variable), living area (dichotomous variable), smoking history (dichotomous variable), drinking history (dichotomous variable), medical insurance status (dichotomous variable), physical status (dichotomous variable), weekly physical exercise time (continuous variable), annual household incomes per capita (continuous variable), and annual neighbourhood incomes per capita (continuous variable). Table 1 shows the summary statistics of variables in the regression models.

Table 1. Summary statistics of variables included in regression analyses.

Variables	Proportion/Mean (SD)
Dependent variables	
CES-D Score (0–60)	7.3 (9.24)
Independent variables	
Neighbourhood social capital	
Neighbourhood social trust	0.78 (0.12)
Neighbourhood social reciprocity	0.48 (0.23)
Neighbourhood social group membership	0.08 (0.15)
$PM_{2.5}$ concentrations ($\mu g/m^3$)	49.27 (19.74)
Control variables	
Gender	
Male	0.48
Female	0.52
Age	44.83 (14.61)
Marital status	
Single, divorced, and widowed	0.19
Married and living with spouse	0.73
Married but living apart from spouse	0.08
Education	
Primary school or below	0.35
High school	0.52
College and above	0.13
Employment	
Employed	0.95
Unemployed	0.05
Hukou status	
Local *hukou*	0.91
Non-local *hukou*	0.09
Living area	
Living in urban neighbourhood	0.39
Living in rural neighbourhood	0.61
Smoking	
Current smoker	0.27
Non-smoker	0.73
Drinking	
Drinker	0.19
Non-drinker	0.81
Medical insurance	
Having medical insurance	0.90
No medical insurance	0.10
Physical health status	
Have a disease	0.11
No disease	0.89
Weekly physical exercise time (minutes)	97.51 (267.95)
Average annual household incomes per household member (Chinese yuan)	17991.68 (202477.08)
Average annual neighbourhood incomes per neighbourhood resident (Chinese yuan)	17814.06 (3.22)
Individual-level social capital	
Trust in neighbours	
Neighbours are extremely/very trustworthy	0.78
Neighbours are somewhat/slightly/not at all trustworthy	0.22
Neighbours are helpful	
Neighbours always/often help each other	0.48
Neighbours sometimes/seldom/never help each other	0.52
Number of types of voluntary groups	0.08 (0.37)

2.3. Statistical Analyses

We examined the effects of cities' $PM_{2.5}$ concentrations on respondents' depressive symptoms in China while we also tested whether neighbourhood social capital moderated the effect of $PM_{2.5}$ concentrations on respondents' depressive symptom, using three-level linear regression analyses. Due to the hierarchical structure of this data set, multilevel models were suitable for this research.

We applied hierarchical liner regression analyses to estimate the effect of $PM_{2.5}$ concentrations on depressive symptoms and the moderating effect of neighbourhood social capital. Models presented here are as follows: a baseline model estimating the effect of controlled variables on depressive symptoms (Model 1), a model estimating the effect of neighbourhood social capital on depressive symptoms (Model 2), a model estimating the effect of $PM_{2.5}$ concentrations on depressive symptoms (Model 3), and a model adding cross-level interaction variables into Model 3 while following the multilevel analysis research (Model 4) [43]. All continuous variables were centred on the grand mean in interaction part. The statistical models were of the following form and the mean value of the variance inflation factor (VIF) was less than 3:

$$\begin{aligned} CES-D_{ihj} = \beta_0 \ &+\beta_1 Neighbourhood\ social\ capital\ indicators_{hj} + \beta_2 PM_{2.5}\ concentrations_j \\ &+\beta_3 Neighbourhood\ social\ capital\ indicators_{hj} \cdot PM_{2.5}\ concentrations_j \\ &+\beta_4 Covariates_h + \beta_5 Covariates_{ihj} + \varepsilon_{ihj} + \mu_{hj} + \varphi_j \end{aligned} \tag{1}$$

where i represents individuals, h represents neighbourhoods, and j represents prefectures. β_0 is the intercept. *Neighbourhood social capital indicators$_{hj}$* represents a vector of neighbourhood-level variables of social capital. *$PM_{2.5}$ concentrations$_j$* represents a vector of prefecture-level variables of *$PM_{2.5}$ concentration.* Social capital indicators$_{hj}$ · $PM_{2.5}$ concentrations$_j$ represents a vector of cross-level interaction effect. *Covariates$_{hj}$* represent a vector of neighbourhood-level covariates. *Covariates$_{ihj}$* represent a vector of individual-level covariates. ε_{ihj}, μ_{hj}, φ_j represent random errors at the individual level, neighbourhood level, and city level, respectively. The interaction effect of interest is expressed in coefficients β_3.

3. Results

Table 2 shows the results of the multilevel linear models on respondents' depressive symptoms. Model 1 included individual-level and neighbourhood-level control variables. Compared with women, men had a lower CES-D score (coefficient = −1.243, standard error = 0.153). In addition, respondents' CES-D scores increased with age (coefficient = 0.040, standard error = 0.005). Married respondents had a lower CES-D score (married and living with a spouse, coefficient = −1.052, standard error = 0.179; married and living apart with a spouse, coefficient = −0.718, standard error = 0.264). Respondents with higher educational attainment had a lower CES-D score, compared with respondents who graduated from primary school or below, (high school, coefficient = −1.065, standard error = 0.153; college and above, coefficient = −1.064, standard error = 0.251). Compared with unemployed respondents, employed respondents have lower CES-D scores (coefficient = −0.558, standard error = 0.265). What's more, respondents with medical insurance had lower CES-D scores (coefficient = −0.850, standard error = 0.203) than those without medical insurance. Respondents with physical diseases have higher CES-D scores (coefficient = 5.897, standard error = 0.196) than those without physical diseases. Furthermore, respondents' CES-D scores decreased with physical exercise time (coefficient = −0.119, standard error = 0.026). Interestingly, respondents' CES-D scores decreased with the logarithm of household incomes and neighbourhood incomes (logarithm of household incomes, coefficient = −0.579, standard error = 0.064; logarithm of neighbourhood incomes, coefficient = −1.150, standard error = 0.311). Lastly, CES-D scores decreased with all three individual-level social capital indicators (neighbours are extremely/very trustworthy, coefficient = −1.736, standard error = 0.148; neighbours always/often help each other, coefficient = −1.133, standard error = 0.131; number of types of voluntary groups, coefficient = −0.261, standard error = 0.131). Model 2 included control variables and neighbourhood social capital indicators.

Table 2. Multilevel liner regression coefficients for the effects of PM2.5 concentrations, social capital, and individual characteristics on depressive symptoms.

Effects and Variables	Model 1 (Baseline)	Model 2	Model 3	Model 4
Fixed part				
Logarithm of $PM_{2.5}$ concentrations			2.167 ** (1.090)	2.670 ** (1.390)
Neighbourhood-level social capital				
Neighbourhood social trust		-4.152 *** (1.443)	-4.247 *** (1.443)	-4.271 *** (1.498)
Neighbourhood social reciprocity		-1.959 *** (0.841)	-1.878 *** (0.854)	-2.083 *** (0.849)
Neighbourhood social group membership		-0.968 ** (0.489)	-0.962 ** (0.481)	-0.146 ** (0.073)
Male (ref: female)	-1.243 *** (0.153)	-1.240 *** (0.153)	-1.240 *** (0.153)	-1.240 *** (0.153)
Age	0.040 *** (0.005)	0.040 *** (0.005)	0.040 *** (0.005)	0.040 *** (0.005)
Marital status and family organization(ref: single, divorced, and widowed)				
Married and living with spouse	-1.053 *** (0.179)	-1.050 *** (0.179)	-1.050 *** (0.179)	-1.050 *** (0.179)
Married but living apart from spouse	-0.718 *** (0.264)	-0.719 *** (0.264)	-0.718 *** (0.264)	-0.719 *** (0.264)
Education (ref: primary school or below)				
High school	-1.065 *** (0.153)	-1.075 *** (0.153)	-1.077 *** (0.153)	-1.074 *** (0.153)
College and above	-1.064 *** (0.251)	-1.078 *** (0.252)	-1.081 *** (0.252)	-1.082 *** (0.252)
Employed (ref: unemployed)	-0.558 ** (0.265)	-0.552** (0.265)	-0.553 ** (0.265)	-0.553 ** (0.265)
Local hukou (ref: non-local hukou)	-0.324 (0.246)	-0.273 (0.247)	-0.273 (0.247)	-0.279 (0.247)
Living in urban neighbourhood(ref: living in rural neighbourhood)	0.108 (0.345)	-0.240 (0.424)	-0.243 (0.424)	-0.223 (0.427)
Current smoking status (ref: non-smoker)	0.100 (0.173)	0.091 (0.173)	0.092 (0.173)	0.088 (0.173)
Current drinking status (ref: non-drinker)	-0.057 (0.171)	-0.056 (0.171)	-0.056 (0.171)	-0.055 (0.171)
Medical insurance (ref: no medical insurance)	-0.850 *** (0.203)	-0.848 *** (0.203)	-0.848 *** (0.203)	-0.846 *** (0.203)
Have a disease (ref: no disease)	5.897 *** (0.196)	5.886 *** (0.196)	5.887 *** (0.196)	5.889 *** (0.196)
Logarithm of physical exercise time	-0.119 *** (0.026)	-0.120 *** (0.026)	-0.120 *** (0.026)	-0.120 *** (0.026)
Logarithm of household incomes per capita	-0.579 *** (0.064)	-0.588 *** (0.064)	-0.588 *** (0.064)	-0.587 *** (0.064)
Logarithm of neighbourhood incomes per capita	-1.150 *** (0.311)	-1.279 *** (0.315)	-1.287 *** (0.316)	-1.206 *** (0.316)
Individual-level social capital				
Neighbours are extremely/very trustworthy (ref: neighbours are somewhat/slightly/not at all trustworthy)	-1.736 *** (0.148)	-1.698 *** (0.149)	-1.698 *** (0.149)	-1.698 *** (0.149)
Neighbours always/often help each other (ref: neighbours sometimes/seldom/never help each other)	-1.133 *** (0.131)	-1.094 *** (0.133)	-1.094 *** (0.133)	-1.095 *** (0.133)
Number of types of voluntary groups	-0.261 ** (0.131)	-0.250 ** (0.125)	-0.250 ** (0.125)	-0.250 ** (0.125)
Cross-level interaction				
Neighbourhood social trust × logarithm of PM2.5 concentrations				-0.846 ** (0.401)
Neighbourhood social reciprocity × logarithm of PM2.5 concentrations				-1.019 *** (0.102)
Neighbourhood social group membership × logarithm of PM2.5 concentrations				-0.924 (0.437)

Table 2. *Cont.*

Effects and Variables	Model 1 (Baseline)	Model 2	Model 3	Model 4
Constant	16.929 *** (0.756)	20.021 *** (1.335)	19.482 *** (2.018)	19.111 *** (1.998)
Random part				
Var (city-level constant)	2.020 ***	2.210 ***	2.190 ***	1.990 ***
Var (neighbourhood-level constant)	5.192 ***	4.922 ***	4.916 ***	4.917 ***
Var (Residual)	70.301 ***	71.299 ***	70.300 ***	70.300 ***
Number of cities	158	158	158	158
Number of neighbourhoods	401	401	401	401
Number of individuals	20,861	20,861	20861	20,861
AIC	148,666.800	148,664.500	148,662.300	148,660.500

** $p < 0.05$, *** $p < 0.01$. All continuous independent variables and covariates were grand-mean centred.

Surprisingly, respondents' CES-D scores decreased with all of neighbourhood social capital indicators (neighbourhood social trust, coefficient = −4.152, standard error = 1.443; neighbourhood social reciprocity, coefficient = −1.959, standard error = 0.841; neighbourhood social group membership, coefficient = −0.968, standard error = 0.489). Model 3 included control variables, social capital indicators, and $PM_{2.5}$ concentrations. The results showed that respondents' CES-D scores increased with logarithm of $PM_{2.5}$ concentrations (coefficient = 2.167, standard error = 1.090) which means $PM_{2.5}$ concentrations had significant negative effect on respondents' mental health. Lastly, cross-level interaction effects were added in Model 4. The effect of $PM_{2.5}$ concentrations on respondents' CES-D scores varied by neighbourhood social capital indicators which means that neighbourhood social capital significantly moderated the relationship between city's $PM_{2.5}$ concentrations and respondents' depressive symptoms.

Figure 3 graphically displays the predicted $PM_{2.5}$ concentrations-depressive symptoms differing by neighbourhood social capital indicators in Model 4, where different neighbourhood social capital indicators are represented by the Lower Quartile (LQ = 25%), the median (MQ = 50%) and higher quartile (HQ = 75%). The result of Model 4 and Figure 3a shows that respondents living in cities with higher concentrations of $PM_{2.5}$ with higher neighbourhood social trust had lower CES-D scores than respondents living in cities with higher concentrations of $PM_{2.5}$ with lower neighbourhood social trust. With the rise of neighbourhood social trust, its moderating effect was strengthened. Figure 3b shows that respondents living in cities with higher concentrations of $PM_{2.5}$ with higher neighbourhood social reciprocity had lower CES-D scores than did respondents living in cities with higher concentrations of $PM_{2.5}$ with lower neighbourhood social reciprocity. With the rise of neighbourhood social reciprocity, its moderating effect was strengthened.

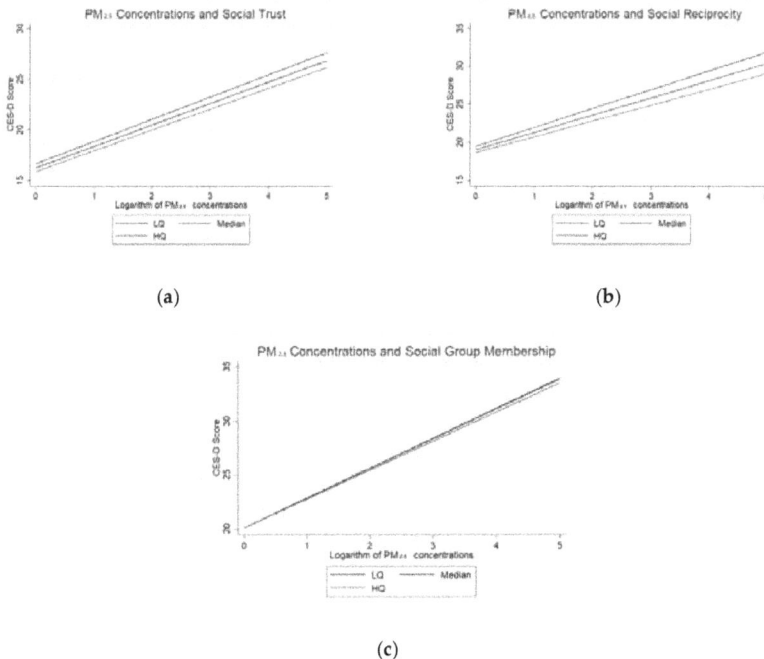

(a)

(b)

(c)

Figure 3. (a) Predicted relationship between $PM_{2.5}$ concentrations and CES-D scores differing by social trust; (b) Predicted relationship between $PM_{2.5}$ concentrations and CES-D scores differing by social reciprocity; (c) Predicted relationship between $PM_{2.5}$ concentrations and CES-D scores differing by social group membership.Lower Quartile (LQ = 25%), the median (MQ = 50%) and higher quartile (HQ = 75%).

Figure 3c shows that respondents living in cities with higher concentrations of $PM_{2.5}$ with higher neighbourhood social group membership had lower CES-D scores than respondents living in cities with higher concentrations of $PM_{2.5}$ with lower neighbourhood social group membership. With the rise of neighbourhood social group membership, its moderating effect was strengthened; however, the moderating effect of the difference of median (MQ = 50%) and higher quartile (HQ = 75%) of neighbourhood social group membership is not obvious.

4. Discussion

In the present study, we investigated the relationship between $PM_{2.5}$ concentrations and depressive symptoms in China. First, as has been demonstrated in previous studies respondents' depressive symptoms increased with cities' $PM_{2.5}$ concentrations [11,13,14,16]. Such a finding may be due to the following reasons: (1) air pollution may increase the risk of cardiovascular diseases, and cardiovascular diseases are closely related to depressive symptoms [17,44]; therefore, residents living in cities with more air pollution are more likely experience both cardiovascular diseases and depressive symptoms, thus continuing a vicious circle. (2) Air pollution may also decrease the frequency of residents' outdoor physical activities, and outdoor physical activities are associated with depressive symptoms [38,39,45,46].

In addition, previous studies we found that neighbourhood social capital was beneficial to residents' health [1,28,30,47]. Social capital can increase access to local services and amenities [28] and provide useful support for residents [28]. Rapid urbanization in China has eroded residents' connection with friends and relatives [48]. Therefore, neighbourhood social capital has become essential to residents' health in China. For this reason, neighbourhood social capital benefits residents' mental health in China.

Most importantly, the statistical significance of the interaction effect indicates that neighbourhood social capital can weaken the negative impact of $PM_{2.5}$ concentrations on depressive symptoms. In other words, neighbourhood social capital exerts a protective effect on the relationship between depressive symptoms and $PM_{2.5}$ concentrations. There are several explanations for the protective effect. First, as for neighbourhood social trust, previous studies have noted that health knowledge spreads faster in high-social-trust neighbourhoods (vs. low) since people are more likely to share health knowledge with others and accept others' advice [28,30,49,50]. Therefore, residents experiencing depressive symptoms caused by $PM_{2.5}$ may acquire useful health knowledge about cardiovascular, nervous, and digestive system disease prevention more easily and can learn how to cope with stressor from lack of sunlight in a high-social-trust neighbourhood. Second, social interactions are more frequent in high-social-trust neighbourhoods, since their residents are more likely to be in contact with those whom they trust [28,30,49,50]. Thus, although $PM_{2.5}$ concentrations may lead to a decrease in face-to-face social contact among neighbours', residents living in high-social-trust neighbourhoods maintain connections with their neighbours through indoor activities (for example, playing card games or Mah-jong). As a result, the negative effect of $PM_{2.5}$ may be weakened by neighbourhood social trust.

In addition, as for neighbourhood social reciprocity, residents can not only obtain emotional support but also can receive material support from neighbours in a high-social-reciprocity neighbourhood [28,30,51–53]. Therefore, residents experiencing depressive symptoms caused by $PM_{2.5}$ may garner emotional comfort by talking to their neighbours about their mood or also acquire useful health knowledge of cardiovascular, nervous and digestive system disease prevention and know how to cope with stressor from lack of sunlight. All this will make residents feel less depressed even while still experiencing $PM_{2.5}$.

Furthermore, as for neighbourhood social group membership, living in a high-social-group-membership neighbourhood, residents are more willing to participate in group activities [28,29]. Although $PM_{2.5}$ may decrease residents' willingness to have physical activities outdoor, living in high-social-group-membership neighbourhood may increase their willingness to have both outdoor and indoor physical activities. Social interactions among group members are more frequent than that among non-group members [28,29]. Even with fewer daily interactions, group members can maintain their interactions by attending routine group activities. In a word, the negative effect of $PM_{2.5}$ is weakened by neighbourhood social group membership.

Lastly, this study also revealed that the protective effect of neighbourhood group membership is relatively weaker than the other two neighbourhood social capital effects, and this may be because this research used the average number of types of voluntary groups within each neighbourhood to measure neighbourhood group membership. However, we failed to measure the strength of the social ties within each group, which means one may be a part of several social groups but fail to garner strong ties in each respective group [28].

From a policy perspective, to decrease depressive symptoms, the government should pay attention to the following three aspects. First, more trees and grass should be planted in cities to increase green space. Green space can benefit residents' health and reduce air pollution [54,55]. In addition, more public medical insurance and subsidies should be provided to low-income residents to weaken the negative effect of environmental hazards on poor residents' health. Finally, neighbourhood social capital should be promoted. Existing studies have found that neighbourhood social capital can be improved by promoting residents' educational attainment; therefore, more funding should be provided to improve national education [56–58].

Despite this study's advantages, some limitations should be noted. First, due to the cross-sectional nature of the data, we were unable to capture time-fixed effects and only calculated average $PM_{2.5}$ concentrations in 2015. Second, we used only $PM_{2.5}$ concentrations as an indicator of air pollution; previous studies have used other air pollution indicators related to ozone, nitrogen dioxide, and sulphur dioxide [11,12]. Lastly, we measured the concentration of $PM_{2.5}$ at the city level; future research should utilize air quality index at a finer geographical level.

5. Conclusions

This study confirms that neighbourhood social capital plays a protective role in the relationship between depressive symptoms and $PM_{2.5}$ concentrations in China. $PM_{2.5}$ concentrations may increase respondents' depressive symptoms, while neighbourhood social capital may decrease respondents' depressive symptoms. Most importantly, neighbourhood social capital weakens the negative influences of $PM_{2.5}$ concentrations on respondents' depressive symptoms. However, the mechanism through which air pollution negatively impacts residents' mental health is still unclear in developing countries. Therefore, further studies are warranted.

Author Contributions: Conceptualization, Ruoyu Wang and Ye Liu; Methodology, Ruoyu Wang; Software, Ruoyu Wang and Penghua Liu; Validation, Ruoyu Wang and Penghua Liu; Formal Analysis, Ruoyu Wang and Penghua Liu; Investigation, Penghua Liu and Hongsheng Chen; Resources, Penghua Liu and Hongsheng Chen; Data Curation, Penghua Liu and Hongsheng Chen; Writing-Original Draft Preparation, Ruoyu Wang and Penghua Liu; Writing-Review & Editing, Ruoyu Wang and Penghua Liu; Visualization, Ruoyu Wang and Penghua Liu; Supervision, Desheng Xue, and Ye Liu; Project Administration, Desheng Xue, and Ye Liu; Funding Acquisition, Desheng Xue, and Ye Liu.

Funding: This research was funded by the National Natural Science Foundation of China (No. 41320104001, No. 41501151) and Innovative R&D Team Introduction Program of Guangdong Province supported this research.

Acknowledgments: The National Natural Science Foundation of China (No. 41320104001, No. 41501151) and Innovative R&D Team Introduction Program of Guangdong Province supported this research. Data analyzed in this paper were collected by the research project "China Labor-force Dynamics Survey (CLDS)" carried out by the Center for Social Survey, Sun Yat-sen University. The authors appreciate the assistance in providing data by the institutes and individuals aforementioned. Please direct your correspondence to liuye25@mail.sysu.edu.cn. All errors remain ours.

Conflicts of Interest: The authors declare no conflict of interest.

References

1. Adjaye-Gbewonyo, K.; Kawachi, I.; Subramanian, S.V.; Avendano, M. High social trust associated with increased depressive symptoms in a longitudinal South African sample. *Soc. Sci. Med.* **2018**, *197*, 127–135. [CrossRef] [PubMed]
2. Murray, C.J.L.; Vos, T.; Lozano, R.; Naghavi, M.; Flaxman, A.D.; Michaud, C.; Ezzati, M.; Shibuya, K.; Salomon, J.A.; Abdalla, S.; et al. Disability-adjusted life years (DALYs) for 291 diseases and injuries in 21 regions, 1990–2010: A systematic analysis for the Global Burden of Disease Study 2010. *Lancet* **2012**, *380*, 2197–2223. [CrossRef]
3. Rumsfeld, J.S.; Ho, P.M. Depression and cardiovascular disease: A call for recognition. *Circulation* **2005**, *111*, 250–253. [CrossRef] [PubMed]
4. Hare, D.L.; Toukhsati, S.R.; Johansson, P.; Jaarsma, T. Depression and cardiovascular disease: A clinical review. *Eur. Heart J.* **2014**, *35*, 1365–1372. [CrossRef] [PubMed]
5. Ruo, B.; Rumsfeld, J.S.; Hlatky, M.A.; Liu, H.; Browner, W.S.; Whooley, M.A. Depressive symptoms and health-related quality of life: The Heart and Soul Study. *JAMA* **2003**, *290*, 215–221. [CrossRef] [PubMed]
6. Stewart, W.F.; Ricci, J.A.; Chee, E.; Hahn, S.R.; Morganstein, D. Cost of lost productive work time among US workers with depression. *JAMA* **2003**, *289*, 3135–3144. [CrossRef] [PubMed]
7. Ekman, M.; Granström, O.; Omerov, S.; Jacob, J.; Landén, M. The societal cost of depression: Evidence from 10,000 Swedish patients in psychiatric care. *J. Affect. Disord.* **2013**, *150*, 790–797. [CrossRef] [PubMed]
8. China Health and Retirement Longitudinal Study 2013. Available online: http://charls.pku.edu.cn/zh-CN (accessed on 25 November 2014).
9. Chan, C.K.; Yao, X. Air pollution in mega cities in China. *Atmosphere. Environ.* **2008**, *42*, 1–42. [CrossRef]
10. Xu, X.; Gao, J.; Dockery, D.W.; Chen, Y. Air pollution and daily mortality in residential areas of Beijing, China. *Arch. Environ. Health: Int. J.* **1994**, *49*, 216–222. [CrossRef] [PubMed]
11. Kioumourtzoglou, M.A.; Power, M.C.; Hart, J.E.; Okereke, O.I.; Coull, B.A.; Laden, F.; Weisskopf, M.G. The association between air pollution and onset of depression among middle-aged and older women. *Am. J. Epidemiol.* **2017**, *185*, 801–809. [CrossRef] [PubMed]
12. Lim, Y.H.; Kim, H.; Kim, J.H.; Bae, S.; Park, H.Y.; Hong, Y.C. Air pollution and symptoms of depression in elderly adults. *Environ. Health Perspect.* **2012**, *120*, 1023. [CrossRef] [PubMed]
13. Cho, J.; Choi, Y.J.; Suh, M.; Sohn, J.; Kim, H.; Cho, S.K.; Ha, K.H.; Kim, C.; Shin, D.C. Air pollution as a risk factor for depressive episode in patients with cardiovascular disease, diabetes mellitus, or asthma. *J. Affect. Disord.* **2014**, *157*, 45–51. [CrossRef] [PubMed]
14. Szyszkowicz, M. Air pollution and emergency department visits for depression in Edmonton, Canada. *Int. J. Occup. Med. Environ. Health* **2007**, *20*, 241–245. [CrossRef] [PubMed]
15. Szyszkowicz, M.; Rowe, B.; Colman, I. Air pollution and daily emergency department visits for depression. *Int. J. Occup. Med. Environ. Health* **2009**, *22*, 355–362. [CrossRef] [PubMed]
16. Wang, Y.; Eliot, M.N.; Koutrakis, P.; Gryparis, A.; Schwartz, J.D.; Coull, B.A.; Mittleman, M.A.; Milberg, W.P.; Lipsitz, L.A.; Wellenius, G.A. Ambient air pollution and depressive symptoms in older adults: Results from the MOBILIZE Boston study. *Environ. Health Perspect.* **2014**, *122*, 553. [CrossRef] [PubMed]
17. Naarding, P.; Schoevers, R.A.; Janzing, J.G.; Jonker, C.; Koudstaal, P.J.; Beekman, A.T. A study on symptom profiles of late-life depression: The influence of vascular, degenerative and inflammatory risk-indicators. *J. Affect. Disord.* **2005**, *88*, 155–162. [CrossRef] [PubMed]
18. Ng, F.; Berk, M.; Dean, O.; Bush, A.I. Oxidative stress in psychiatric disorders: Evidence base and therapeutic implications. *Int. J. Neuropsychopharmacol.* **2008**, *40*, 851–876. [CrossRef] [PubMed]
19. Kent, S.T.; McClure, L.A.; Crosson, W.L.; Arnett, D.K.; Wadley, V.G.; Sathiakumar, N. Effect of sunlight exposure on cognitive function among depressed and non-depressed participants: A REGARDS cross-sectional study. *Environ. Health* **2009**, *8*, 34. [CrossRef] [PubMed]
20. Benedetti, F.; Colombo, C.; Barbini, B.; Campori, E.; Smeraldi, E. Morning sunlight reduces length of hospitalization in bipolar depression. *J. Affect. Disord.* **2001**, *62*, 221–223. [CrossRef]
21. Zhang, X.; Zhang, X.; Chen, X. Happiness in the air: How does a dirty sky affect mental health and subjective well-being? *J. Environ. Econ. Manag.* **2017**, *85*, 81–94. [CrossRef] [PubMed]
22. Li, F.; Liu, Y.; Lü, J.; Liang, L.; Harmer, P. Ambient air pollution in China poses a multifaceted health threat to outdoor physical activity. *J. Epidemiol. Commun. Health* **2015**, *69*, 201. [CrossRef] [PubMed]

23. Giles, L.V.; Koehle, M.S. The Health Effects of Exercising in Air Pollution. *Sports Med.* **2014**, *44*, 223–249. [CrossRef] [PubMed]
24. Wen, X.J.; Balluz, L.S.; Shire, J.D.; Mokdad, A.H.; Kohl, H.W.I. Association of self-reported leisure-time physical inactivity with particulate matter 2.5 air pollution. *J. Environ. Health* **2009**, *72*, 40–44. [PubMed]
25. Lin, G.Z.; Li, L.; Song, Y.F.; Zhou, Y.X.; Shen, S.Q.; Ou, C.Q. The impact of ambient air pollution on suicide mortality: A case-crossover study in Guangzhou, China. *Environ. Health* **2016**, *15*, 90. [CrossRef] [PubMed]
26. Kim, C.; Jung, S.H.; Kang, D.R.; Kim, H.C.; Moon, K.T.; Hur, N.W.; Shin, D.C.; Suh, I. Ambient particulate matter as a risk factor for suicide. *Am. J. Psychiatry* **2010**, *167*, 1100–1107. [CrossRef] [PubMed]
27. Wong, C.M.; Yang, L.; Thach, T.Q.; Chau, P.Y.; Chan, K.P.; Thomas, G.N.; Lam, T.H.; Wong, T.W.; Hedley, A.J.; Peiris, J.S. Modification by Influenza on Health Effects of Air Pollution in Hong Kong. *Environ. Health Perspect.* **2009**, *117*, 248–253. [CrossRef] [PubMed]
28. Kawachi, I.; Kennedy, B.P.; Glass, R. Social capital and self-rated health: A contextual analysis. *Am. J. Public Health* **1999**, *89*, 1187–1193. [CrossRef] [PubMed]
29. Niedzwiedz, C.L.; Richardson, E.A.; Tunstall, H.; Shortt, N.K.; Mitchell, R.J.; Pearce, J.R. The relationship between wealth and loneliness among older people across Europe: Is social participation protective? *Prev. Med.* **2016**, *91*, 24–31. [CrossRef] [PubMed]
30. Boneham, M.A.; Sixsmith, J.A. The voices of older women in a disadvantaged community: Issues of health and social capital. *Soc. Sci. Med.* **2006**, *62*, 269–279. [CrossRef] [PubMed]
31. Feng, Z.; Vlachantoni, A.; Liu, X.; Jones, K. Social trust, interpersonal trust and self-rated health in China: A multi-level study. *Int. J. Equity Health* **2016**, *15*, 180. [CrossRef] [PubMed]
32. Lindström, M.; Moghaddassi, M.; Merlo, J. Individual self-reported health, social participation and neighbourhood: A multilevel analysis in Malmo, Sweden. *Prev. Med.* **2004**, *39*, 135–141. [CrossRef] [PubMed]
33. Abbott, S.; Freeth, D. Social capital and health: Starting to make sense of the role of generalized trust and reciprocity. *J. Health Psychol.* **2014**, *13*, 874–883. [CrossRef] [PubMed]
34. Weitzman, E.R.; Kawachi, I. Giving means receiving: The protective effect of social capital on binge drinking on college campuses. *Am. J. Public Health* **2000**, *90*, 1936–1939. [PubMed]
35. Holtgrave, D.R.; Crosby, R. Is Social Capital a Protective Factor Against Obesity and Diabetes? Findings from an Exploratory Study. *Ann. Epidemiol.* **2006**, *16*, 406–408. [CrossRef] [PubMed]
36. Evans, G.W.; Kutcher, R. Loosening the link between childhood poverty and adolescent smoking and obesity: The protective effects of social capital. *Psychol. Sci.* **2011**, *22*, 3–7. [CrossRef] [PubMed]
37. Murayama, H.; Yu, N.; Matsuo, E.; Nishi, M.; Taniguchi, Y.; Fujiwara, Y.; Shinkai, S. Are neighborhood bonding and bridging social capital protective against depressive mood in old age? A multilevel analysis in Japan. *Soc. Sci. Med.* **2015**, *124*, 171–179. [CrossRef] [PubMed]
38. Ard, K.; Colen, C.; Becerra, M.; Velez, T. Two mechanisms: The role of social capital and industrial pollution exposure in explaining racial disparities in self-rated health. *Int. J. Environ. Res. Public Health* **2016**, *13*, 1025. [CrossRef] [PubMed]
39. China Labor Force Dynamics Survey. Secondary China Labor Force Dynamics Survey 2017. Available online: http://css.sysu.edu.cn/Data (accessed on 27 November 2017).
40. Airborne Fine Particulate Matter and Air Quality Index. Secondary Airborne Fine Particulate Matter and Air Quality Index 2016. Available online: http://www.pm25.in/ (accessed on 25 June 2013).
41. Liu, P.; Yao, Y.; Liang, H.; Liang, Z.; Zhang, Y.; Wang, H. Analyzing spatiotemporal distribution of PM(2.5) in China by integrating Kalman filter and multilevel clustering. *J. Geo-Inform. Sci.* **2017**, *19*, 475–485. [CrossRef]
42. Kalman, R.E.; Bucy, R.S. New results in linear filtering and prediction theory. *J. Basic Eng.* **1961**, *83*, 95–108. [CrossRef]
43. Davidian, M. Hierarchical linear models: Applications and data analysis methods. *J. Am. Stat. Assoc.* **2003**, *98*, 767–768. [CrossRef]
44. Miller, K.A.; Siscovick, D.S.; Sheppard, L.; Shepherd, K.; Sullivan, J.H.; Anderson, G.L.; Kaufman, J.D. Long-term exposure to air pollution and incidence of cardiovascular events in women. *N. Engl. J. Med.* **2007**, *356*, 447–458. [CrossRef] [PubMed]
45. Chen, H.; Liu, Y.; Zhu, Z.; Li, Z. Does where you live matter to your health? Investigating factors that influence the self-rated health of urban and rural Chinese residents: Evidence drawn from Chinese General Social Survey data. *Health Qual. Life Outcomes* **2017**, *15*, 78. [CrossRef] [PubMed]

46. Lü, J.; Liang, L.; Feng, Y.; Li, R.; Liu, Y. Air pollution exposure and physical activity in China: Current knowledge, public health implications, and future research needs. *Int. J. Environ. Res. Public Health* **2015**, *12*, 14887–14897. [CrossRef] [PubMed]

47. Phelan, J.C.; Link, B.G.; Tehranifar, P. Social conditions as fundamental causes of health inequalities: Theory, evidence, and policy implications. *J. Health Soc. Behav.* **2010**, *51*, S28–S40. [CrossRef] [PubMed]

48. Wang, Y.P.; Wang, Y.; Jiansheng, W.U. Urbanization and informal development in China: Urban villages in Shenzhen. *Int. J. Urban Reg. Res.* **2009**, *33*, 957–973. [CrossRef]

49. Cornwell, B.; Laumann, E.O. The health benefits of network growth: New evidence from a national survey of older adults. *Soc. Sci. Med.* **2015**, *125*, 94–106. [CrossRef] [PubMed]

50. Feng, Z.; Jones, K.; Wang, W.W. An exploratory discrete-time multilevel analysis of the effect of social support on the survival of elderly people in China. *Soc. Sci. Med.* **2015**, *130*, 181–189. [CrossRef] [PubMed]

51. Fuhrer, R.; Stansfeld, S.A. How gender affects patterns of social relations and their impact on health: A comparison of one or multiple sources of support from "close persons". *Soc. Sci. Med.* **2010**, *54*, 811–825. [CrossRef]

52. Kumar, S.; Calvo, R.; Avendano, M.; Sivaramakrishnan, K.; Berkman, L.F. Social support, volunteering and health around the world: Cross-national evidence from 139 countries. *Soc. Sci. Med.* **2012**, *74*, 696–706. [CrossRef] [PubMed]

53. Lee, S.Y.D.; Arozullah, A.M.; Cho, Y.I. Health literacy, social support, and health: A research agenda. *Soc. Sci. Med.* **2004**, *58*, 1309–1321. [CrossRef]

54. Gascon, M.; Triguero-Mas, M.; Martínez, D.; Dadvand, P.; Rojas-Rueda, D.; Plasència, A.; Nieuwenhuijsen, M.J. Residential green spaces and mortality: A systematic review. *Environ. Int.* **2016**, *86*, 60–67. [CrossRef] [PubMed]

55. Liu, X.; Liang, X.; Li, X.; Xu, X.; Ou, J.; Chen, Y.; Li, S.; Wang, S.; Pei, F. A future land use simulation model (FLUS) for simulating multiple land use scenarios by coupling human and natural effects. *Landsc. Urban Plan.* **2017**, *168*, 94–116. [CrossRef]

56. Torpe, L.; Lolle, H. Identifying social trust in cross-country analysis: Do we really measure the same? *Soc. Indic. Res.* **2011**, *103*, 481–500. [CrossRef]

57. Liu, Y.; Xu, W. Destination choices of permanent and temporary migrants in China, 1985–2005. *Popul. Space Place* **2017**, *23*, e1963. [CrossRef]

58. Liu, Y.; Zhang, F.; Wu, F.; Liu, Y.; Li, Z. The subjective wellbeing of migrants in Guangzhou, China: The impacts of the social and physical environment. *Cities* **2017**, *60*, 333–342. [CrossRef]

International Journal of
*Environmental Research
and Public Health*

MDPI

Article

Depression in Intimate Partner Violence Victims in Slovenia: A Crippling Pattern of Factors Identified in Family Practice Attendees

Nena Kopčavar Guček and Polona Selič *

Department of Family Medicine, Faculty of Medicine, University of Ljubljana, Poljanski nasip 58,
1000 Ljubljana, Slovenia; nenagucek@gmail.com
* Correspondence: polona.selic@siol.net; Tel.: +386-31-379-707

Received: 22 November 2017; Accepted: 24 January 2018; Published: 26 January 2018

Abstract: This multi-centre cross-sectional study explored associations between prevalence of depression and exposure to intimate partner violence (IPV) at any time in patients' adult life in 471 participants of a previous IPV study. In 2016, 174 interviews were performed, using the Short Form Domestic Violence Exposure Questionnaire, the Zung Scale and questions about behavioural patterns of exposure to IPV. Family doctors reviewed patients' medical charts for period from 2012 to 2016, using the Domestic Violence Exposure Medical Chart Check List, for conditions which persisted for at least three years. Depression was found to be associated with any exposure to IPV in adult life and was more likely to affect women. In multivariable logistic regression modelling, factors associated with self-rated depression were identified ($p < 0.05$). Exposure to emotional and physical violence was identified as a risk factor in the first model, explaining 23% of the variance. The second model explained 66% of the variance; past divorce, dysfunctional family relationships and a history of incapacity to work increased the likelihood of depression in patients. Family doctors should consider IPV exposure when detecting depression, since lifetime IPV exposure was found to be 40.4% and 36.9% of depressed revealed it.

Keywords: intimate partner violence; self-rated depression; mental health; family practice; sick leave; incapacity to work

1. Introduction

Violence and mental health problems and their possible association and coexistence, have been extensively researched and published [1]. According to WHO, domestic violence should be prioritized as a public health problem, due to its prevalence as well as its consequences [2]. The associations between mental health problems and exposure to IPV in the victims and also in the perpetrators, are complex [1,3]. While substance abuse co-morbidity and a past history of violence are considered the strongest predictors of future violence, current evidence is not enough to suggest that severe mental illness can independently predict violent behaviour [4–9].

Exposure to psychological abuse was found to be more strongly associated with the prevalence of depression, anxiety, somatization, experiencing suicidal thoughts and post-traumatic stress disorder than with other types of IPV [3,8–10]. Women are more likely to become victims of IPV and twice as likely to become depressed [11,12]. The gender difference in the frequency of depression can be accounted for by women experiencing greater poverty, differing social roles and sex discrimination, more negative life events and violence and abuse [13]. Evidence indicates that partner abuse may contribute to depression [13]. Female victims who had suffered IPV in the past year were found to have their relative risk of depression increased by 3.26 compared to non-abused women [14].

A meta-analysis on the prevalence of mental health problems in women with a history of IPV in 1999 found that just under half of the abused women had clinical depression [15].

Abundant literature on the consequences of IPV on health has warned against undetected depression in IPV victims [16]; IPV-related depression was first discussed as a relevant health-related problem in family medicine several years ago [9], yet has never been studied as such in Slovenia. Given that the data for past IPV abuse had already been re-evaluated, showing a prevalence of exposure to all types of IPV of 17% in family medicine clinics attendees [17] and a prevalence of psychological IPV alone of approximately 10%, it was of the utmost importance for Slovenian family medicine to focus on the health-related outcomes of exposure to IPV. This study therefore aimed to explore the associations between the prevalence of depression in patients who were exposed to IPV at any time in their adult lives and those who were not; to identify the health consequences and other patient characteristics associated with exposure to IPV; and to examine whether there were any specifically gender-related issues in family clinic attendees. Since prior results showed that current health status was not associated with the current psychological IPV exposure [9], this study design covered a longer time frame when assessing IPV-related health conditions and the prevalence of depression, providing a list of health conditions that need in-depth exploration towards possible IPV exposure in family medicine attendees, i.e., chronic pain syndrome, incapacity to work, muscle inflammations (myalgia, muscle soreness or musculoskeletal pain), gastrointestinal disorders, irregularities in bowel functioning, reduced physical functioning, gynaecological disorders, genital tract infections, the state of depression and/or generalized anxiety disorder, eating and sleeping disorders, phobias and panic attacks, low self-esteem and psychosomatic disorders.

2. Materials and Methods

2.1. Participants and Procedure

Sixty-four family doctors (FDs), i.e., physicians who have finished four years of specialized training and who had already taken part in the 2012 IPV prevalence re-evaluation study [17] and participated in the 2013 Psychological IPV study [9], were invited to participate in this follow-up study in September 2016.

The 2013 study [9] was also multi-centre and cross-sectional, recruiting 960 family practice attendees aged 18 years and above, without dementia or even mild cognitive impairment, who were willing to participate. In 689 interviews with currently- or previously-partnered patients, the short form of A Domestic Violence Exposure Questionnaire and additional questions about behavioural patterns of exposure to psychological abuse in the past year were given. The FDs reviewed the medical charts of 470 patients who had been in an intimate relationship in the previous five-year period. Since we aimed to analyse solely psychological abuse, patients who had experienced psychological IPV during their lifetime but not in the past year, as well as victims of multiple types of abuse, were excluded and offered help and assistance. Using the Domestic Violence Exposure Medical Chart Check List, data on the patients' lives, physical, sexual and reproductive and psychological health status, as well as sick leave, hospitalisation, visits to family clinics and referrals to other clinical specialists in the past year were gathered. In multivariable logistic regression modelling the factors associated with past year psychological IPV exposure were identified, with unemployment or working part-time ($p = 0.001$), a college degree ($p = 0.038$), an intimate relationship of six years or more ($p = 0.048$) and a history of disputes in the intimate relationship ($p < 0.001$) increasing the odds of emotional abuse and explaining 41% of the variance. In females, unemployment ($p = 0.002$) and a history of disputes in the intimate relationship ($p < 0.001$) explained 43% of the variance.

In present study, FDs were asked if they had attended the "Recognizing and Treating Victims of Domestic Violence in Health Care Settings: Guidelines and Training for Health Professionals" training program for improving the competences of health professionals in recognizing and responding to victims of domestic violence. This education and training for health professionals took place between

September 2015 and March 2016 at the Medical Chamber of Slovenia and was funded by Norway Grants [18]. Thirty-seven responded affirmatively and were considered properly empowered and competent to meet this study's requirements. The FDs were provided with written instructions on their approach to the patients, including an invitation letter to the patients, eligibility criteria and on data collection, i.e., the depression section of the Composite International Diagnostic Interview (CIDI) [16], semi-structured interview forms and medical charts review forms; they were already familiar with the latter since the same forms were used in the 2013 Psychological IPV study [9].

2.1.1. The First Phase: Identification of Patients

The participating FDs were asked to identify the patients from the 2013 study and invite them to contact the designated family clinic and make an appointment. The first phase of data collection was planned for January to March 2017 but was actually carried out during March and April, when the FDs, who work in family practices all over the country, interviewed 283 patients (of 471 invited) who attended and were willing to participate. Other eligibility criteria for this phase were age and the absence of dementia or even mild cognitive impairment. In the invitation letter to patients, the aim of the study was explained as being a follow-up on people's wellbeing after a four-year gap and the subjects were also informed that participation was not obligatory. Those willing to participate made appointments and were scheduled for an interview at the patients' convenience within two to seven weeks. This phase of data collection ended on 30 April 2017.

2.1.2. The Second Phase: Interviewing the Patients

The FDs explained to the participants (283 of 471 invited (the response rate being 60.1%)) the associations between exposure to IPV and quality of life. Prior to this phase, the FDs revised the elements to be covered in the semi-structured interviews. After the introduction to the IPV topic, 246 patients signed a written consent form, while 37 decided against participation and were not interviewed any further. The interviewing phase lasted until the end of April; during the 246 interviews, 72 patients revealed they had not lived together with an intimate partner in their adulthood or at least in the past decade They explained that they were living either with their parents, their children, both or alone. Those who had never shared a household with an intimate partner were excluded from further analysis. During the interview, a further 13 patients declared they were not willing to complete the self-assessment scale for depression. As the study aimed to analyse associations between self-rated depression and possible IPV exposure, patients without a correctly completed depression scale sheet were excluded from the analysis regardless of their IPV exposure; however, they were offered help and assistance, similarly to the previous study. After the self-administration of the SDS, a nurse calculated the total score and the FDs, who were instructed to evaluate the screening results, interviewed patients who scored above 30 points, using the depression section of the CIDI [16].

2.1.3. The Third Phase: Auditing Patients' Medical Charts

The auditing of patients' medical charts was carried out using The Domestic Violence Exposure Medical Chart Check List [19,20], described in detail elsewhere. All medical conditions were checked from 2012, as requested in the 2013 IPV study, up to the present time, covering the whole five years. Only those persisting at least for three years, either three consecutive years or three times in a specified period of inspection, were marked. Given that the FDs participated in the 2013 study, they were familiar with the list. However, they were provided with additional clarifications regarding several variables which were found to be ambiguous in the previous study [9]: incapacity for work was defined as the inability of the victim, due to their exposure to IPV, to perform the normal duties of work in their job or at the post occupied; reduced physical functioning was clarified strictly as a functional and consequently psychosocial impairment i.e., not attending any recreational physical activities, neglecting one's hobbies and interests, and/or omitting attendance at social functions and duties; a dysfunctional family was marked as a family in which conflict, lack of empathy and support, misbehaviour, and/or child neglect or abuse on the part of individual parents occurred; and sick leave assessment was taken into account only as absence from work measured in days due to the consequences of family violence.

2.1.4. Participants: Study Sample

In total, the medical charts of 161 participants, male and female, who declared that they had been living with their intimate partners at least during the past decade were reviewed. The drop-out data of the recruitment process are presented in Figure 1.

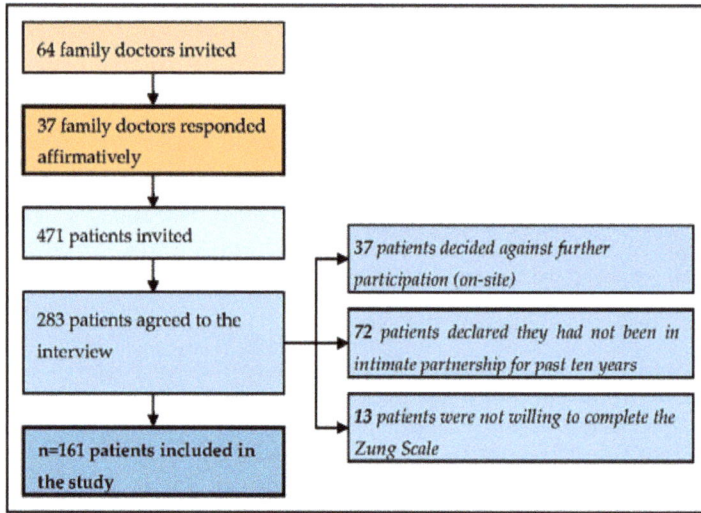

Figure 1. The drop-out data.

The mean age of all the participants (*n* = 161) was 51.1 ± 16.8 (range 22–92 years), while the mean age of the partner (of those patients who were married or were living in an intimate relationship, *n* = 115) was 49.7 ± 16.1 (range 20–93 years). The mean value for Zung SDS was 33.2 ± 9.3. In total, 65 (40.4%) participants were exposed to IPV at any time in their adult life.

The National Medical Ethics Committee of the Republic of Slovenia approved the protocol of the study (document number 77/01/15 from 19 January 2014).

2.2. Measures

The short form of A Domestic Violence Exposure Questionnaire was developed in previous studies in Slovenian primary care [19–21] and used as an exploration tool, consisting of specific questions with comprehensive, behaviourally-defined descriptions of interpersonal violence events which were to be considered within the patients' whole adult life. An adult was defined by the right to vote at the age of 18 years, regardless of financial, emotional or other types of autonomy in the individual. The questions were worded as follows: *Do you feel safe at home? Do you feel accepted, respected and loved in your intimate relationship? Have you been humiliated, subjected to threats, insult or intimidation, or in any way emotionally affected by your intimate partner? Does your partner talk down to you? Has he/she demeaned or insulted you or made you feel ashamed? Has he/she screamed or cursed at you? Has he/she threatened you with physical harm?* The patients were not asked to specify whether this had happened in the previous year and/or any number of years preceding the survey, or with a current or former intimate partner; for the aim of this study, it was important for patients to disclose if they had ever been exposed to IPV in their adult lives. However, the FDs noted answers referring to events in the past year.

Additionally, the patients were asked whether, aside from the aforementioned behaviours, they had endured any patterns of physical activity that had harmed or might have harmed them, e.g., slapping, kicking, pushing, being forced into sexual activities. The interviews were closed by asking *If such a thing has happened, have you been thinking about doing something about it? Do you want me to help?* The remaining data collected were on gender, age, number of children, marital status and number of divorces and place of residence.

The Zung Self-Rating Depression Scale (SDS) [22] was used to assess depression. In the 2013 study [9], depression was evaluated using the using the depression section of the Composite International Diagnostic Interview, which provided psychiatric diagnoses based on symptoms experienced in the last six months, according to the ICD-10 criteria. Several risk factors, intrinsic either to the individual or to the social context, were included. However, the quality control of these interviews was not strictly performed and questions about interviewers' competencies related to bias appeared. The Zung scale has been widely used in Slovenia [23] so it was decided on, regardless of better measures being available, since we wanted to be able to compare results with previous findings. The Zung scale consists of 20 statements, 10 of them reversed and each one is scored from 1 to 4, making the range of the completed scale from 20 to 80. All the participants completed the Zung SDS and a depressive score was calculated and interpreted as follows: less than 39 points, no depression; between 40 and 49 points, marginal depression; and above 50 points, depression was indicated. Cronbach's alpha was 0.79. The cut-off score of 40 points was calculated based on the CIDI, which was performed after the SDS administration. The FDs categorised the CIDI results as depression indicated and depression not indicated.

After the interviews took place, the FDs reviewed the patients' medical charts, replicating the procedure from previous IPV studies in Slovenian family medicine [9,17,21], focusing on the health-related associations of exposure to IPV and data on the patients' wider life context at both the personal and relationship level, as reported by the patient in previous years. Health consequences were later categorized into three groups: physical, sexual and reproductive and psychological; this procedure has been described in depth elsewhere [9].

Aiming to follow up the 2013 study results, sick leave (in episodes and days), hospitalization (in episodes and days), visits to the family clinic and referrals to other clinical specialists in the past year were reviewed and are presented in Table 1. 'Frequent' was defined as within the top 10 percentile in a time frame of one year for each characteristic. The utilization of healthcare services was additionally analysed concerning self-rated depression.

Table 1. The utilization of healthcare services concerning IPV exposure and self-rated depression in the last 12 months.

	Top 10 Percentiles	Violence Any Time in Adult Life		*p	Violence Any Time in Adult Life (Female Patients)		p*	Self-Rated Depression		p*	Self-Rated Depression (Female Patients)		p*
		No n = 96 (%)	Yes n = 65 (%)		No n = 65 (%)	Yes n = 58 (%)		No n = 127 (%)	Yes n = 34 (%)		No n = 92 (%)	Yes n = 31 (%)	
Sick leave (episodes)	3 or more	16 (16.7)	9 (13.8)	0.822	12 (18.5)	6 (10.3)	0.297	18 (14.2)	7 (20.6)	0.250	11 (12.0)	7 (22.6)	0.125
Sick leave (days)	46 or more	5 (5.2)	5 (7.7)	0.504	5 (7.7)	4 (6.9)	1.000	5 (3.9)	5 (14.7)	0.020	4 (4.3)	5 (16.1)	0.032
Hospitalization (episodes)	1 or more	7 (7.3)	7 (10.8)	0.551	6 (9.2)	6 (10.3)	1.000	8 (6.3)	6 (17.6)	0.080	7 (7.6)	5 (16.1)	0.294
Hospitalization (days)	1 or more	6 (6.3)	7 (10.8)	0.382	5 (7.7)	6 (10.3)	0.756	7 (5.5)	6 (17.6)	0.027	6 (6.5)	5 (16.1)	0.134
Visits to family clinic	16 or more	13 (13.5)	6 (9.2)	0.460	10 (15.4)	6 (10.3)	0.426	10 (7.9)	9 (26.5)	0.007	7 (7.6)	9 (29.0)	0.005
Referrals to other clinical specialists	5 or more	11 (11.5)	9 (13.8)	0.809	8 (12.3)	8 (13.8)	1.000	14 (11.0)	6 (17.6)	0.382	12 (13.0)	4 (12.9)	1.000

* Fisher's exact test.

2.3. Data Analysis

In the data analysis, frequencies and percentages were used to describe the main characteristics of this study sample (Table 2).

Table 2. Demographics characteristics and IPV exposure of the sample.

	Total *n* (%)	IPV Exposure *n* (%)		*p*
	n = 161 (%)	*n* = 96 (%)	Yes *n* = 65 (%)	
Gender				
Male	38 (23.6)	31 (32.3)	7 (10.8)	0.002 *
Female	123 (76.4)	65 (67.7)	58 (89.2)	
Age				
29 years or less	18 (11.2)	11 (11.5)	7 (10.8)	0.990 #
30–59 years	91 (56.6)	54 (56.3)	37 (56.9)	
60 years or more	52 (32.3)	31 (32.3)	21 (32.3)	
Married or living in an intimate relationship				
No	46 (28.6)	23 (24.0)	23 (35.4)	0.155 *
Yes	115 (71.5)	73 (76.0)	42 (64.6)	
Divorce in the past				
No	125 (77.6)	82 (85.4)	43 (66.2)	0.006 *
Yes	36 (22.4)	14 (14.6)	22 (33.8)	
Age of the children				
Less than 15 years	69 (42.9)	49 (51.0)	20 (30.8)	0.015 *
15 years or more	92 (57.1)	47 (49.0)	45 (69.2)	
Living in urban setting				
No	27 (16.8)	16 (16.7)	11 (16.9)	1.000*
Yes	135 (83.2)	80 (83.3)	54 (83.1)	
Employment status				
Unemployed or working part time	13 (8.1)	4 (4.2)	9 (13.8)	0.086 #
Regularly employed	96 (59.6)	60 (62.5)	36 (55.4)	
Retired	52 (32.3)	32 (33.3)	20 (30.8)	
Level of education				
Elementary school	15 (9.3)	6 (6.3)	9 (13.8)	0.223 #
High school	71 (44.1)	42 (43.8)	29 (44.6)	
College degree or more	75 (46.6)	48 (50.0)	27 (41.5)	
Monthly income per family member				
Less than 900 EUR	84 (52.2)	43 (44.8)	41 (63.1)	0.025 *
900 EUR or more	77 (47.8)	53 (55.2)	24 (36.9)	

* Fisher's exact test, # chi-square test.

In power calculation, for the design effect the calculation equation $1 + (m - 1) \times \rho$ given by Donner et al. [24] was used. The intra-cluster correlation coefficient (ρ) was taken from a large sample study by Smeeth et al. [25] for cluster randomized trials and surveys at the primary care clinic level (a maximum value observed of 0.05). The mean cluster size (m) was 4.4 ± 1.7 patients (range 3–8) and the design effect was equal to 1.17. Consequently, the effective sample size was calculated at $n = 137$ (the actual sample size divided by the design effect). Considering sample size tables by Hsieh [26], a total of 137 cases was determined to have 70% power to detect a significant association for logistic regression (using alpha of 0.05 and a medium odds ratio of about 2.5 to 1 [27]).

The patients were compared with regard to self-rated depression ($\leq$40 (no depression) and >40 (borderline depression or depression)) and also the IPV exposure during their adult life ('an IPV experience' vs. 'never'); see Tables 3 and 4. Multivariable logistic regression was used to model demographic characteristics in association with self-rated depression (Table 5) and afterwards another multivariable logistic modelling was performed, also including variables derived from medical chart analysis, i.e., divorce in the past (yes/no); history of conflict or dispute in the family (yes/no); dysfunctional family relations (yes/no); physical condition: incapacity to work; physical condition: reduced physical functioning; sexual and reproductive consequences of IPV exposure: gynaecological disorders; and exposure to IPV at any time (yes/no). The results were presented by adjusted odds ratios with 95% confidence intervals. Statistical analysis was performed by IBM SPSS 20.0 software (IBM Corp., Armonk, NY, USA) and $p < 0.05$ was set as the level of statistical significance.

Table 3. Self-rated depression in patients exposed to IPV according to their SDS score.

	≤40 (No Depression) $n = 127$ (%)	>40 (Borderline Depression or Depression) $n = 34$ (%)	p *
Emotional violence in the last 12 months ($n = 46$)	29 (22.8)	17 (50.0)	0.003
Physical violence at any time in adult life ($n = 28$)	15 (11.8)	13 (38.2)	0.001
Emotional violence at any time in adult life ($n = 61$)	39 (30.7)	22 (64.7)	0.001
Violence at any time in adult life ($n = 65$)	41 (32.3)	24 (70.6)	<0.001

* Fisher's exact test, SDS = Zung Self-Rated Depression Scale.

Table 4. Patients' medical charts review summary taking into account self-rated depression and IPV exposure in patients.

	Depression According to SDS n (%)		cOR (95%CI), p *	IPV Exposure n (%)		cOR (95%CI), p *
	No Depression n = 127 (%)	Depression n = 34 (%)		No IPV n = 96	IPV n = 65	
Physical status						
Injuries: head, thoracic and abdominal area (n = 13)	11 (8.9)	2 (6.1)	0.66 (0.14–3.13), 1.000	8 (8.3)	5 (7.7)	0.92 (0.29–2.94), 1.000
Scratches and bruises (n = 30)	24 (19.4)	6 (17.6)	0.92 (0.34–2.47), 1.000	15 (15.6)	15 (23.1)	1.62 (0.73–3.60), 0.302
Chronic pain syndrome (n = 132)	100 (81.5)	31 (91.2)	2.79 (0.79–9.83), 0.204	73 (76.0)	59 (90.8)	3.10 (1.18–8.11), 0.021
Incapacity to work (n = 47)	26 (21.0)	21 (61.8)	6.28 (2.78–14.18), <0.001	22 (22.9)	25 (38.5)	2.10 (1.05–4.19), 0.036
Muscle inflammation (n = 93)	67 (54.0)	26 (76.5)	2.91 (1.22–6.92), 0.019	48 (50.0)	45 (69.2)	2.25 (1.16–4.36), 0.022
Bone fractures (n = 31)	28 (22.8)	3 (8.8)	0.34 (0.10–1.20), 0.089	18 (18.8)	13 (20.0)	1.08 (0.49–2.40), 0.842
Gastrointestinal disorders (n = 99)	74 (60.2)	25 (73.5)	1.99 (0.86–4.61), 0.167	50 (52.1)	49 (75.4)	2.82 (1.41–5.63), 0.003
Irregularities in bowel functioning (n = 66)	47 (38.2)	19 (55.9)	2.16 (0.98–4.64), 0.078	28 (29.2)	38 (58.5)	3.42 (1.76–6.62), <0.001
Lacerations and cuts (n = 19)	15 (12.2)	4 (11.8)	0.99 (0.31–3.22), 1.000	11 (11.5)	8 (12.3)	1.08 (0.41–2.86), 1.000
Eye injuries (n = 3)	3 (2.4)	0 (0.0)	0.52 (0.03–10.22), 1.000	1 (1.0)	2 (3.1)	3.02 (0.27–33.97), 0.566
Reduced physical functioning (n = 39)	20 (16.1)	19 (55.9)	6.78 (2.96–15.52), <0.001	16 (16.7)	23 (35.4)	2.74 (1.31–5.74), 0.009
Sexual and reproductive status						
Gynaecological disorders (n = 55)	37 (29.8)	18 (52.9)	2.73 (1.26–5.94), 0.015	25 (26.0)	30 (46.2)	2.43 (1.25–4.75), 0.011
Infertility (n = 8)	7 (5.6)	1 (2.9)	0.52 (0.06–4.37), 1.000	6 (6.3)	2 (3.1)	0.48 (0.09–2.44), 0.476
Genital tract infections (n = 70)	51 (41.1)	19 (55.9)	1.89 (0.88–4.05), 0.172	32 (33.3)	38 (58.5)	2.81 (1.47–5.40), 0.002
Complicated pregnancies/spontaneous abortions (n = 25)	18 (14.6)	7 (20.6)	1.71 (0.65–4.51), 0.430	11 (11.5)	14 (21.5)	2.32 (0.99–5.44), 0.119
Sexual dysfunctions (n = 1)	0 (0.0)	1 (2.9)	11.42 (0.45–286.69), 0.215	1 (1.0)	0 (0.0)	0.49 (0.02–12.12), 1.000
Sexually transmitted diseases, including HIV/AIDS (n = 10)	7 (5.6)	3 (8.8)	1.65 (0.41–6.79), 0.449	6 (6.3)	4 (6.2)	0.98 (0.27–3.63), 1.000
Unplanned/unwanted pregnancies (n = 1)	0 (0.0)	1 (2.9)	11.42 (0.45–286.69), 0.215	0 (0.0)	1 (1.5)	4.49 (0.18–111.90), 0.404
Psychological and behavioural status						
Abuse of alcohol and drugs (n = 5)	5 (4.0)	0 (0.0)	0.32 (0.02–5.98), 0.586	2 (2.1)	3 (4.6)	2.72 (0.37–14.00), 0.394
Depression and/or generalized anxiety disorder (n = 79)	46 (37.4)	33 (97.1)	58.11 (7.69–438.97), <0.001	34 (35.4)	45 (69.2)	4.10 (2.09–8.04), <0.001
Eating and sleeping disorders (n = 81)	50 (41.0)	31 (91.2)	15.91 (4.62–54.85), <0.001	35 (36.5)	46 (70.8)	4.22 (2.14–8.30), <0.001
Feelings of shame and guilt (n = 12)	4 (3.3)	8 (23.8)	9.46 (2.65–33.78), 0.001	4 (4.2)	8 (12.3)	3.23 (0.93–11.21), 0.069
Phobias and panic attacks (n = 30)	18 (14.6)	12 (35.3)	3.30 (1.40–7.86), 0.012	12 (12.5)	18 (27.7)	2.68 (1.19–6.04), 0.022
Physical inactivity (n = 50)	31 (25.2)	19 (55.9)	3.92 (1.78–8.63), 0.001	30 (31.3)	20 (30.8)	0.98 (0.49–1.93), 1.000
Low self-esteem (n = 33)	13 (10.6)	20 (58.8)	12.53 (5.13–30.56), <0.001	14 (14.6)	19 (29.9)	2.42 (1.11–5.27), 0.029
Post-traumatic stress disorder (n = 28)	18 (14.6)	10 (29.4)	2.52 (0.94–6.15), 0.073	15 (15.6)	13 (20.0)	1.38 (0.61–3.14), 0.528
Psychosomatic disorder (n = 107)	74 (60.2)	33 (97.1)	23.64 (3.13–178.26), <0.001	57 (59.4)	50 (76.9)	2.28 (1.13–4.62), 0.027
Smoking (n = 30)	28 (22.8)	2 (5.9)	0.22 (0.05–0.98), 0.027	15 (15.6)	15 (23.1)	1.62 (0.73–3.60), 0.302
Suicidal behaviour and self-harm (n = 2)	1 (0.8)	1 (2.9)	3.82 (0.23–62.68), 0.387	1 (1.0)	1 (1.5)	1.48 (0.91–24.17), 1.000
Unsafe sexual behaviour (n = 1)	1 (0.8)	0 (0.0)	1.22 (0.05–30.67), 1.000	1 (1.0)	0 (0.0)	0.49 (0.02–12.12), 1.000

* Fisher's exact test, SDS = Zung Self-Rated Depression Scale, IPV = intimate partner violence, cOR = crude odds ratio, 95% CI = 95% confidence interval.

Table 5. Associations between self-rated depression and bio-psycho-social characteristics in patients.

	Depression According to SDS *n* (%)		aOR	(95% CI)		*p*
	No *n* = 127	Yes *n* = 34				
Gender						
Male	35 (27.6)	3 (8.8)	Ref	-	-	-
Female	92 (72.4)	31 (91.2)	2.17	0.56	8.45	0.264
Age						
29 years or less	14 (11.0)	4 (11.8)	Ref	-	-	-
30–59 years	73 (57.5)	18 (52.9)	0.62	0.13	2.97	0.549
30–60 years or more	40 (31.5)	12 (35.3)	0.91	0.10	8.56	0.935
Married or living in an intimate relationship						
No	33 (26.0)	13 (38.2)	Ref	-	-	-
30–Yes	94 (74.0)	21 (61.8)	0.99	0.36	2.70	0.982
Divorce in the past						
No	105 (82.7)	20 (58.8)	Ref	-	-	-
30–Yes	22 (17.3)	14 (41.2)	2.50	0.86	7.24	0.092
Age of the children						
Less than 15 years	58 (45.7)	11 (32.4)	Ref	-	-	-
30–15 years or more	69 (54.3)	23 (67.6)	0.83	0.22	3.20	0.791
Living in urban setting						
No	21 (16.5)	6 (17.6)	Ref	-	-	-
30–Yes	106 (83.5)	28 (82.4)	1.13	0.34	3.78	0.846
Employment status						
Unemployed or working part time	8 (6.3)	5 (14.7)	Ref	-	-	-
Regularly employed	78 (61.4)	18 (52.9)	0.58	0.13	2.53	0.467
30–Retired	41 (32.3)	11 (32.4)	0.57	0.08	4.28	0.587
Level of education						
Elementary school	10 (7.9)	5 (14.7)	Ref	-	-	-
High school	58 (45.7)	13 (38.2)	0.55	0.11	2.72	0.461
30–College degree or more	59 (46.5)	16 (47.1)	0.74	0.13	4.24	0.733
Monthly income per family member						
Less than 900 EUR	63 (49.6)	21 (61.8)	Ref	-	-	-
30–900 EUR or more	64 (50.4)	13 (38.2)	0.84	0.29	2.47	0.752
Exposure to violence, any time						
No violence	86 (67.7)	10 (29.4)	Ref	-	-	-
Emotional	28 (22.0)	11 (32.4)	3.04	1.12	8.27	0.029
30–Physical	13 (10.2)	13 (38.2)	4.69	1.42	15.49	0.011

chi-square = 25.711; df = 14; *p* = 0.028; Nagelkerke R^2 = 0.229, SDS = Zung Self-Rated Depression Scale, aOR = adjusted odds ratio, 95% CI = 95% confidence interval.

3. Results

3.1. Demographic Characteristics of Patients

The demographic characteristics of the sample are presented in Table 2. Of the sample, 40.4% (*n* = 65) people had been exposed to IPV at some time in their adulthood (58 women and 7 men).

3.2. Self-Rated Depression in Patients Exposed to IPV

Of the patients exposed to IPV (*n* = 65), 24 (36.9%) scored >40 on the SDS. Overall the participants scored 33.2 ± 9.3 on the SDS; minimum 20, maximum 57. More results on self-rated depression in IPV patients are described in Table 3.

3.3. Frequent Use of Health Care Services by Participating Patients during the Past Year

There were no statistically significant differences found between people exposed to any type of IPV and those who did not report such experiences (Table 1).

3.4. Patients' Medical Charts Review—Summary

The summary of the patients' medical charts is presented in Table 4. Statistically significant differences in physical status were identified between the patients who were—according to the SDS—depressed and those who were not, with regard to incapacity to work ($p < 0.001$), muscle inflammations, i.e., myalgia, muscle soreness or musculoskeletal pain ($p < 0.001$) and reduced physical functioning ($p < 0.019$). Gynaecological disorders were significantly more frequent in the depressed patients ($p = 0.015$), as well as several items in the psychological behavioural status, including the state of depression and/or generalized anxiety disorder ($p < 0.001$), eating and sleeping disorders ($p < 0.001$), feelings of shame and guilt ($p = 0.001$), phobias and panic attacks ($p = 0.012$), physical inactivity ($p = 0.001$), low self-esteem ($p < 0.001$), psychosomatic disorders ($p < 0.001$) and smoking ($p = 0.027$), all assessed or diagnosed by practicing FDs from the year prior to the 2013 survey [9] (March 2012) until April or at the latest December 2016, if persistent for at least three years, i.e., three consecutive years or three times during this period.

According to the 40.4% IPV exposure at any time in patients in Table 4, statistically significant differences in physical status were identified between the patients who were exposed to IPV and those who were not, with regard to chronic pain syndrome ($p = 0.021$), incapacity to work ($p = 0.036$), muscle inflammations (myalgia, muscle soreness or musculoskeletal pain; $p = 0.022$), gastrointestinal disorders ($p = 0.003$), irregularities in bowel functioning ($p < 0.001$) and reduced physical functioning ($p = 0.009$). Gynaecological disorders were significantly more frequent in IPV exposed patients ($p = 0.011$), as were genital tract infections ($p = 0.002$), as well as several items in the psychological behavioural status, including the state of depression and/or generalized anxiety disorder ($p < 0.001$), eating and sleeping disorders ($p < 0.001$), phobias and panic attacks ($p = 0.022$), low self-esteem ($p < 0.001$) and psychosomatic disorders ($p < 0.001$); all were assessed or diagnosed by practising FDs.

3.5. Associations between Self-Rated Depression and Bio-Psycho-Social Characteristics in Patients: Logistic Regression Modelling

In the regression modelling process, the associations between self-rated depression and the characteristics of patients were explored. Exposure to emotional (aOR 3.04, 95% CI 1.12–8.27, $p = 0.029$) and to physical violence (aOR 4.69, 95% CI 1.42–15.49, $p = 0.011$) were identified as risk factors, explaining 23% of the variance (Nagelkerke $R^2 = 0.229$, $p = 0.028$). The results are presented in Table 5.

When analysing only female patients, there was almost the same division of explained variance (Nagelkerke $R^2 = 0.227$, $p = 0.084$) and the two risk factors identified (exposure to emotional (aOR 3.41, 95% CI 1.13–10.37, $p = 0.030$) and to physical violence (aOR 5.14, 95% CI 1.46–18.09, $p = 0.011$)).

3.6. Associations between Self-Rated Depression, the Bio-Psycho-Social Characteristics and Medical Charts Review Summary Patients: Logistic Regression Modelling

Additionally, the logistic regression modelling was broadened, also taking into account data gathered in the process of patients' medical charts review and performed to explore self-rated depression and its associations in all patients and separately in female patients. In females, divorce in the past (aOR 8.80, 95% CI 1.21–63.91, $p = 0.032$), dysfunctional family relationships (aOR 25.30, 95% CI 3.12–205.37, $p = 0.02$) and a history of incapacity to work identified by the GP in the patient's medical chart review (aOR 22.16, 95% CI 2.71–181.08, $p = 0.004$) increased the odds of depression in female patients, with regression modelling explaining 65% of the variance (Nagelkerke $R^2 = 0.655$, $p < 0.001$). The model which included both female and male patients explained 66% of the variance (Nagelkerke $R^2 = 0.657$, $p < 0.001$) with divorce in the past (aOR 6.11, 95% CI 1.12–33.48, $p = 0.037$), dysfunctional family relationships (aOR 19.29, 95% CI 3.13–118.87, $p = 0.001$) and a history of incapacity to work identified by the GP in the patient's medical chart review (aOR 18.94, 95% CI 3.26–109.97, $p = 0.001$) increased the odds of depression in patients.

4. Discussion

The prevalence of depression was found to be associated with exposure to IPV anytime in the adult life and women were found to be more likely to be victimized (Tables 3 and 5). The physical health consequences associated with exposure to IPV were incapacity to work, reduced physical functioning and muscle inflammation, while the psychological consequences included smoking, panic attacks and anxiety, eating disorders, feelings of shame and guilt and low self-esteem (Tables 1, 3 and 4).

Depression in family practice attendees in Slovenia was found to be significantly associated with female gender, chronic conditions, chronic pain and age above 45 years in a study by Klemenc Ketiš [23]. The Self-Rated Depression Scale was used in that study as well as in the present one and the difference in the overall rate of depression detected in the participants (15.2% vs. 27.8%) could be attributed to the different mean age of the participants (44.2 vs. 51.1 years). Nearly 37% of all the participants in our study who had been exposed to violence at any time in their adult lives fulfilled the criteria for depression according to the SDS (Table 3), compared to 21.7% of depressive patients with chronic conditions and to 23.2% of those with chronic pain [23]. Past exposure to IPV was not only associated with depression but also with suicide attempts, according to a meta-analysis by Devries et al. [28]. Since IPV has been claimed to trigger a whole range of depressive symptoms, this might explain why the rate of depression in the participants exposed to IPV was higher than in those with chronic conditions and chronic pain [23]. Another study by Selič et al. in 2013, in the same environment, detected an 89.5% rate of depression and general anxiety disorder in the participants exposed to IPV [9]. A more recent publication by the same author introduced rates of depression in elderly people in Slovenia, reporting depression to be significantly associated with female gender, age, years of education, smoking, number of concurrent diseases and self-rated health regarding heart, blood vessels, musculoskeletal system and mental health but comparably to our results (Table 2)—not with marital status or alcohol consumption [29]. As the mean age of the participants was 75.1 ± 6.2 years, the empty nest stage of the family cycle, the remoteness of possible exposure to IPV and possibly living alone may diminish the influence of partner interaction on depression. This finding is congruent with the research of Bonomi et al. [30,31], which found that recent exposure to physical and/or sexual IPV increased the risk of depressive and severe depressive symptoms. The authors emphasized that more pronounced adverse health effects were identified in women with recent (vs. remote) exposure to IPV. Our findings are comparable with those in IPV survivors [30,31]: reported exposure to partner violence and physical violence in both the past year and also at any time in adult life were found to be a statistically significant risk factors for current, self-rated depression (Table 3) and also for generalized anxiety disorder and the prevalence of depression (Table 4). Dysfunctional family relationships were found to be significantly associated with the prevalence of depression in those exposed to IPV ($p = 0.001$); according to Felitti, inquiry into depression, past abuse and past or present dysfunctional family life should be added to the current clinical evaluation [32]. Poor family functioning has long been recognized as a factor associated with IPV [33,34], either as the cause or the consequence of the IPV, similar to our findings with IPV related depression in all patients and also separately in female patients ($p = 0.002$). The association between IPV, poor family functioning and divorce in the past ($p = 0.032$) and consequent depression in participants in this study is concordant with that described as complex and multidirectional by others [1,11,35].

Several authors report that victims of violence seek medical assistance more often than violence free individuals [11,36,37], which was not confirmed in our study (Table 1). However, the utilisation of healthcare services, i.e., sick leave (46 or more days) and visits to the family clinic (16 or more) were more frequent in depressed attendees, all patients and females only ($p_{all} = 0.020$, $p_{females} = 0.032$; $p_{all} = 0.007$, $p_{females} = 0.005$, respectively); while hospitalization (1 or more days) was more likely to be used by this study sample ($p = 0.027$) but by not females separately (Table 1). Evidencing the underlying cause of a specific sick leave as a consequence of IPV may be challenging and unclear and could contribute to the non-significant difference in the sick leave of individuals exposed to IPV versus those who were not (Table 1). In extreme cases, the IPV and consequent health issues do not only result in incapacity for work but also cause permanent loss of working capability, debilitation, deterioration of quality of life and premature death [6,10,38]. In accordance with these data, incapacity to work and reduced physical functioning in our study were the most significant physical consequences associated with self-rated depression (Table 4) and also with IPV exposure in adult life (Table 4). A longitudinal study on a representative sample of Slovenian family medicine patients [39] brought good insight into the quality of life related factors. Since the sample was representative, it is important to be aware of factors significantly and consistently associated with a better mental component score of quality of life, i.e., social support, satisfactory circumstances in patients' household (absence of violence) and absence of anxiety.

4.1. Limitations of the Study

The sample in this study could not be considered as representative of a family medicine attendees' population [40], since there were more women (76.4% vs. 54.8%) and the level of education was much higher (46.6% vs. 11.3% people with college degree or above, 9.3% vs. 41.0% with elementary school) but the mean age was almost the same (51.1 ± 16.8 years vs. 51.7 ± 19.0 years). Yet there are not so many noticeable differences in comparison with the 2013 study [9], i.e., in the present study there were also more women (76.4% vs. 61.3%) and the mean age of this study's participants was slightly older (51.1 ± 16.8 years vs. 47.4 ± 16.1 years), while the level of education appears similar (46.6% vs. 45.7% people with college degree or above, 9.3% vs. 11.1% with elementary school). Given the females prevailing in the sample (76.4%), the findings of this study should be used with caution, without emphasizing gender specificities.

The opportunistic sampling method, i.e., the sample size of 161 out of 471 invited patients, may not have reflected the diversity of the Slovenian community and might have affected the IPV exposure prevalence, which was found to be 40.4% for the whole adult life. It is therefore of the utmost importance that another study of lifetime IPV exposure in Slovenian family medicine attendees be performed soon, regardless of possible mental health consequences as a special focus of interest.

Reliance on self-reported data and social desirability bias may have both impacted on results, given that IPV is still considered as a private and/or too sensitive an affair for many people to intervene in Slovenia [41]. The burden of somatic co-morbidity was shown to be smaller than the impact of psychosocial determinants in a study on a representative sample of family practice attendees [16], emphasizing the social stigma of mood disorders. As explained in the Method section, there were 13 participants unwilling to complete the SDS. Given that the detection and successful treatment of depression and anxiety have already been recognized as facilitating factors with a potential to lead to improved quality of life in family medicine attendees some time ago [40] and consequently FDs were instructed to be alert for the early onset of these conditions, apparently more efforts should be make in awareness-raising in FDs.

The cross-sectional design, known as inherently limited, together with part of the obtained data being self-reported, could have raised hindrances about the potential for method variance to account for this study's findings. Since the social environment related phenomenon, i.e., the IPV exposure being studied could have been assessed only by asking patients to report their experience or perception, it was of great advantage that the research design incorporated medical records to assemble the exact health-related data (Table 4). Except for the self-reported data on IPV and depression, there were several characteristics followed through the longer period of time (from March 2012 to December 2016 at the latest) reviewed by the FDs in the patients' medical history, e.g., results of the Composite International Diagnostic Interview conducted in 2013 study [9], physical, sexual and reproductive IPV related health consequences and psychological and behavioural status, which in our opinion alleviated the potential effects of method variance. The partition of explained variance in multivariable modelling grew evidently higher when these data was added ($R^2 = 0.657$, $p < 0.001$; $R^2_{(females)} = 0.655$, $p_{(females)} < 0.001$) to illustrate the associations between self-rated depression, bio-psycho-social characteristics of patients and their medical charts review summary.

4.2. Implications for Future Research and Family Practice

Although only exposure to emotional and physical IPV, both being self-reported, identified as being associated with depression (Table 5), the data was collected in a large number of settings, which hopefully reduced the non-response bias. However, future research should be more focused in following up the initial cohort of patients, given that for this study purpose the cohort was identified with less difficulty than expected, while the FDs had undergone a thorough capacity building process in the field of IPV exposure related concepts, both practical and theoretical.

For the first time in Slovenian family medicine, a clear indication of the association of IPV exposure with depression was found, in spite of the barriers on the side of Slovenian FDs detected in previous studies [41].

We were unable to avoid a high drop-out rate compared to the previous study [9] (Figure 1), so rethinking this future study plan, together with the need to evaluate the 40.4% lifetime IPV exposure prevalence, would be beneficial.

5. Conclusions

The prevalence of depression in patients who were exposed to IPV at any time in their adult life was significantly more likely than in non-IPV exposed patients ($p < 0.001$); in those who were exposed to IPV, almost 37% (36.92% ($n = 24$)) were diagnosed as depressed. For this reason, FDs detecting depression in their patients will be advised to explore the possibility of exposure to IPV as a contributing factor and vice versa. Practising FDs should be aware of the overlapping effect of symptoms, e.g., incapacity to work, depression and anxiety in patients. Therefore, those lacking satisfactory household circumstances, regardless of a clear IPV exposure revealed, should be recognized and empowered. The assessment of two out of five adult family medicine attendees who might have been exposed to IPV strengthens this recommendation.

Author Contributions: Polona Selič conceived the study, carried out the coordination and drafted the manuscript. Nena Kopčavar Guček participated in the execution of the study and interpretation and helped to draft the manuscript. Both authors read and approved the final manuscript.

Conflicts of Interest: The authors declare no conflict of interest.

References

1. Varshney, M.; Mahapatra, A.; Krishnan, V.; Gupta, R.; Deb, K.S. Violence and mental illness: What is the true story? *J. Epidemiol. Community Health* **2016**, *70*, 223–225. [CrossRef] [PubMed]
2. Preamble to the Constitution of the World Health Organization as Adopted by the International Health Conference, New York, 19–22 June 1946. Available online: https://hero.epa.gov/hero/index.cfm/reference/details/reference_id/80385 (accessed on 31 January 2016).
3. Reisenhofer, S.; Taft, A. Women's journey to safety—The Transtheoretical model in clinical practice when working with women experiencing Intimate Partner Violence: A scientific review and clinical guidance. *Patient Educ. Couns.* **2013**, *93*, 536–548. [CrossRef] [PubMed]
4. Juninger, J.; McGuire, L. Psychotic motivation and the paradox of current research on serious mental illness and rates of violence. *Schizophr. Bull.* **2004**, *30*, 21–30. [CrossRef]
5. Kishor, S.; Johnson, K. Profiling Domestic Violence: A Multi-Country Study. Calverton, MD, ORC Macro. 2004, pp. 27–43. Available online: http://dhsprogram.com/pubs/pdf/OD31/OD31.pdf (accessed on 31 March 2014).
6. Coker, A.L.; Smith, P.H.; Bethea, L.; Kong, M.R.; McKeown, R.E. Physical health consequences of physical and psychological intimate partner violence. *Arch. Fam. Med.* **2000**, *9*, 451–457. [CrossRef] [PubMed]

7. Elbogen, E.B.; Johnson, S.C. The intricate link between violence and mental disorder: Results from the National Epidemiologic Survey on Alcohol and Related Conditions. *Arch. Gen. Psychiatry* **2009**, *66*, 152–161. [CrossRef] [PubMed]
8. Amore, M.; Menchetti, M.; Tonti, C.; Scarlatti, F.; Lundgren, E.; Esposito, W.; Berardi, D. Predictors of violent behavior among acute psychiatric patients: Clinical study. *Psychiatry Clin. Neurosci.* **2008**, *62*, 247–255. [CrossRef] [PubMed]
9. Selic, P.; Svab, I.; Kopcavar Gucek, N. A cross-sectional study identifying the pattern of factors related to psychological intimate partner violence exposure in Slovenian family practice attendees: What hurt them the most. *BMC Public Health* **2014**, *14*, 223. [CrossRef] [PubMed]
10. Coker, A.L.; Davis, K.E.; Arias, I.; Desai, S.; Sanderson, M.; Brandt, H.M.; Smith, P.H. Physical and mental health effects of intimate partner violence for men and women. *Am. J. Prev. Med.* **2002**, *23*, 260–268. [CrossRef]
11. World Health Organization. *Global and Regional Estimates of Violence against Women: Prevalence and Health Effects of Intimate Partner Violence and Non-Partner Sexual Violence*; World Health Organization (WHO): Geneva, Switzerland, 2013; ISBN 978 92 4 156462.
12. Hyde, J.S.; Mezulis, A.H.; Abramson, L.Y. The ABCs of depression: Integrating affective, biological, and cognitive models to explain the emergence of the gender difference in depression. *Psychol. Rev.* **2008**, *115*, 291–313. [CrossRef] [PubMed]
13. Astbury, J.; Cabral, M. *Women's Mental Health: An Evidence Based Review*; WHO/MSD/MDP/00.1; World Health Organization: Geneva, Switzerland, 2000; pp. 31–89.
14. Bonomi, A.E.; Anderson, M.L.; Reid, R.J.; Rivara, F.P.; Carrell, D.; Thompson, R.S. Medical and Psychosocial Diagnoses in Women with a History of Intimate Partner Violence. *Arch. Intern. Med.* **2009**, *169*, 1692–1697. [CrossRef] [PubMed]
15. Golding, J. Intimate partner violence as a risk factor for mental disorders: A meta-analysis. *J. Fam. Violence* **1999**, *14*, 99–132. [CrossRef]
16. Selič, P.; Svab, I.; Rifel, J.; Pavlič, D.R.; Cerne, A.; King, M.; Nazareth, I. The pattern of physical comorbidity and the psychosocial determinants of depression: A prospective cohort study on a representative sample of family practice attendees in Slovenia. *Ment. Health Fam. Med.* **2011**, *8*, 147–155. [PubMed]
17. Selic, P.; Svab, I.; Kopcavar Gucek, N. How many Slovenian family practice attendees are victims of intimate partner violence? *A re-evaluation cross-sectional study report. BMC Public Health* **2013**, *13*, 703. [CrossRef]
18. POND Project. Recognizing and Treating Victims of Domestic Violence in Slovenia. Available online: http://eeagrants.org/News/2015/Recognising-and-treating-victims-of-domestic-violence-in-Slovenia (accessed on 31 July 2017).
19. Selic, P.; Pesjak, K.; Kersnik, J. The prevalence of exposure to domestic violence and the factors associated with co-occurrence of psychological and physical violence exposure: A sample from primary care patients. *BMC Public Health* **2011**, *11*, 621. [CrossRef] [PubMed]
20. Kopčavar Guček, N.; Švab, I.; Selič, P. The prevalence of domestic violence in primary care patients in Slovenia in a five-year period (2005–2009). *Croat. Med. J.* **2011**, *52*, 728–734. [CrossRef] [PubMed]
21. Selič, P.; Pesjak, K.; Kopčavar Guček, N.; Kersnik, J. Dejavniki, ki povečujejo možnost nasilja v družini in iskanje pomoči pri zdravniku družinske medicine. Pilotna študija o nasilju v družini. (Factors that increase likelihood of violence in the family and seeking help from the family practitioner. Pilot study about violence in the family). *Zdr. Vestnik* **2008**, *77*, 505–510.
22. Zung, W.W. A self-rating depression scale. *Arch. Gen. Psychiatry* **1965**, *12*, 63–70. [CrossRef] [PubMed]
23. Klemenc Ketiš, Z. The presence of anxiety and depression in the adult population of family practice patients with chronic diseases. *Zdr. Varst.* **2009**, *48*, 170–176.
24. Donner, A.; Birkett, N.; Buck, C. Randomization by cluster. Sample size requirements and analysis. *Am. J. Epidemiol.* **1981**, *114*, 906–914. [CrossRef] [PubMed]
25. Smeeth, L.; Ng, E.S. Intraclass correlation coefficients for cluster randomized trials in primary care: Data from the MRC trial of the assessment and management of older people in the community. *Control Clin. Trials* **2002**, *23*, 409–421. [CrossRef]
26. Hsieh, F.Y. Sample size tables for logistic regression. *Stat. Med.* **1989**, *8*, 795–802. [CrossRef] [PubMed]
27. Rosenthal, J.A. Qualitative descriptors of strength of association and effect size. *J. Soc. Serv. Res.* **1996**, *21*, 37–59. [CrossRef]

28. Devries, K.M.; Mak, J.; Bacchus, L.J.; Child, J.C.; Falder, G.; Petzold, M.; Astbury, J.; Watts, C.H. Intimate Partner Violence and Incident Depressive Symptoms and Suicide Attempts: A Systematic Review of Longitudinal Studies. *PLoS Med.* **2013**, *10*, e1001439. [CrossRef] [PubMed]

29. Selič, P. Cross-sectional study exploring factors associated with depression in elderly living at home. In *Continuous Development of Nursing in Society and its Contribution to Health Promotion = Proceedings of Lectures with Peer Review*; Skela-Savič, B., Hvalič Touzery, S., Eds.; Angela Boškin Faculty of Health Care: Jesenice, Slovenia, 2017; pp. 189–196.

30. Umubyeyi, A.; Mogren, I.; Ntaganira, J.; Krantz, G. Intimate partner violence and its contribution to mental disorders in men and women in the post genocide Rwanda: Findings from a population based study. *BMC Psychiatry* **2014**, *14*, 315. [CrossRef] [PubMed]

31. Bonomi, A.E.; Thompson, R.S.; Anderson, M.; Reid, R.J.; Carell, D.; Dimer, J.A.; Rivara, F.P. Intimate Partner Violence and Women's Physical, Mental, and Social Functioning. *Am. J. Prev. Med.* **2006**, *30*, 447–540. [CrossRef] [PubMed]

32. Felitti, V.J. Childhood sexual abuse, depression, and family dysfunction in adult obese patients: A case control study. *South. Med. J.* **1993**, *86*, 732–736. [CrossRef] [PubMed]

33. Heise, L.; Garcia Moreno, C. Violence by intimate partners. In *World Report on Violence and Health*; Krug, E.G., Dahlberg, L.L., Mercy, J.A., Zwi, A.B., Lozano, R., Eds.; World Health Organization: Geneva, Switzerland, 2002; pp. 87–122.

34. Black, D.A.; Heyman, R.E.; Smith Slep, A.M. Risk factors for child physical abuse. *Aggress. Violent Behav.* **2001**, *6*, 121–188. [CrossRef]

35. Mechanic, M.B.; Weaver, T.L.; Resick, P.A. Mental health consequences of intimate partner abuse: A multidimensional assessment of four different forms of abuse. *Violence Women* **2008**, *14*, 634–654. [CrossRef] [PubMed]

36. Dahlberg, L.L.; Krug, E.G. Violence—A global preventive problem. In *World Report on Violence and Health*; Krug, E.G., Dahlberg, L.L., Mercy, J.A., Zwi, A.B., Lozano, R., Eds.; World Health Organization: Geneva, Switzerland, 2002; pp. 1–2.

37. WHA49.25. Prevention of Violence: A Public Health Priority. Forty-Ninth World Health Assembly Geneva 1996. Volume 3. Available online: http://www.who.int/violence_injury_prevention/resources/publications/en/WHA4925_eng.pdf (accessed on 6 June 2013).

38. Wagner, P.; Mongan, P. Validating the concept of abuse: Women's perceptions of defining behaviors and the effects of emotional abuse on health indicators. *Arch. Fam. Med.* **1998**, *7*, 25–29. [CrossRef] [PubMed]

39. Cerne, A.; Svab, I.; Kersnik, J.; Selic, P. Did past economic prosperity affect the health related quality of life predictors? A longitudinal study on a representative sample of Slovenian family medicine patients. *BMC Public Health* **2013**, *13*, 1160. [CrossRef] [PubMed]

40. Svab, I.; Petek Ster, M.; Kersnik, J.; Zivcec Kalan, G.; Car, J. A cross sectional study of performance of Slovene general practitioners (English summary). *Zdr. Varst.* **2005**, *44*, 183–192.

41. Kopčavar Guček, N.; Petek, D.; Švab, I.; Selič, P. Barriers to Screening and Possibilities for Active Detection of Family Medicine Attendees Exposed to Intimate Partner Violence. *Zdr. Varst.* **2015**, *55*, 11–20. [CrossRef] [PubMed]

International Journal of
Environmental Research and Public Health

MDPI

Article

Exploring the Impacts of Housing Condition on Migrants' Mental Health in Nanxiang, Shanghai: A Structural Equation Modelling Approach

Yang Xiao [1], Siyu Miao [1], Chinmoy Sarkar [2,*], Huizhi Geng [1] and Yi Lu [3]

[1] College of Architecture and Urban Planning, Tongji University, 1239 Siping Road, Shanghai 200092, China; yxiao@tongji.edu.cn (Y.X.); msy0207@163.com (S.M.); genghuizhi@163.com (H.G.)
[2] Healthy High Density Cities Lab, HKUrbanLab, The University of Hong Kong, Knowles Building, Pokfulam Road, Hong Kong, China
[3] Department of Architecture and Civil Engineering, City University of Hong Kong, Hong Kong, China; yilu24@cityu.edu.hk
* Correspondence: csarkar@hku.hk

Received: 2 December 2017; Accepted: 24 January 2018; Published: 29 January 2018

Abstract: Although rapid urbanization and associated rural-to-urban migration has brought in enormous economic benefits in Chinese cities, one of the negative externalities include adverse effects upon the migrant workers' mental health. The links between housing conditions and mental health are well-established in healthy city and community planning scholarship. Nonetheless, there has thusfar been no Chinese study deciphering the links between housing conditions and mental health accounting for macro-level community environments, and no study has previously examined the nature of the relationships in locals and migrants. To overcome this research gap, we hypothesized that housing conditions may have a direct and indirect effects upon mental which may be mediated by neighbourhood satisfaction. We tested this hypothesis with the help of a household survey of 368 adult participants in Nanxiang Town, Shanghai, employing a structural equation modeling approach. Our results point to the differential pathways via which housing conditions effect mental health in locals and migrants. For locals, housing conditions have direct effects on mental health, while as for migrants, housing conditions have indirect effects on mental health, mediated via neighborhood satisfaction. Our findings have significant policy implications on building an inclusive and harmonious society. Upstream-level community interventions in the form of sustainable planning and designing of migrant neighborhoods can promote sense of community, social capital and support, thereby improving mental health and overall mental capital of Chinese cities.

Keywords: housing condition; neighbourhood satisfaction; mental health; migrants; structural equation modelling (SEM); Shanghai

1. Introduction

After three decades of reforms and policy of liberalization, China is currently undergoing rapid urban transformation. China's urbanization rate rose from 17.92% in 1978 to 57.35% in 2016. Nonetheless, the data shows that the urbanization rate of the household population is only about 36%, and there is still a gap relative to the average level achieved in developed countries [1]. Associated with this urbanization phenomenon are large-scale infrastructure projects related to the development of new cities and towns as well as retrofitting of existing ones and a palpable demographic shifts through the influx of rural migrant workers to the cities. The United Nations Development Program (UNDP) projects that China will have an additional 310 million new urban dwellers in the coming 20 years, with the total urban population reaching 1 billion. In other words, one in eight urban residential dwellers on Earth are likely to live in Chinese cities [2].

As Chinese urbanism begins to shape its residents' lives, it brings in several opportunities and challenges to their health and wellbeing [3,4]. Studies in the West have shown that urbanization brings together human capital, innovations, and special political, cultural, economic, and educational opportunities, while it has a negative externalities in the form of health costs at both the psychological and physiological levels [5]. Higher degrees of urbanization have been related with higher the incidences of mental illness [6], and improving mental health and capital of populations and cities is of primary importance [7]. In one of the seminal studies, Bairoch reported significantly higher mortality rates in European cities as compared to adjoining rural areas [8]. Recently, Oculicz-Kozaryn reported quantitative evidence indicating that the largest and densest cities are least happy [9]. Chinese cities are experiencing considerable stress on account of massive rural-to-urban migration and the mental disorders constitute a major burden, accounting for 9.5% of all disability-adjusted life years and 23.6% of all years lived with disability [10,11]. Mental disorders constitute a major health risk for migrants workers [12]. With the prospect of new urban life, migrants are faced with the challenge of establishing themselves in a new social landscape, often quite different from their previous established support networks in terms of sense of belonging and social capital. More importantly, re-settlement, employment, and the pressure to support dependent family members are all associated with a high psychological burden [13] and cause psycho-social stress, anxiety, and depression and other mental disorders [14]. Furthermore, *hukou* or the individual household registration is non-transferable in China, which means that once a migrant moves from rural to urban areas, their entitlements to social and medical benefits are significantly diminished [15], which has negative effects upon quality of life, physical, and mental health.

Housing is the fundamental component of urban environment lying along the path from urban environment to mental health; high-density congested housing has a negative impacts on mental health, both on account of poor physical environment [16] and due to its potential to interfere with and weaken the family's social support system [17]. In extreme cases, it can even affect the next generation of children's psychological health [18]. The World Health Organization (WHO) in 1977 put forward the "everyone is healthy" concept. The WHO in 2009 stressed that healthy cities should be able to encourage, meet, and support the health, well-being, security, social relations, accessibility, mobility, self-esteem, and cultural identity of all social and age groups.

The pathways from housing environment to individual mental health linkages are very complex [19]. In general, the main effects are often mediated by the social environment, physical space, access to social and health services provided by the city, as well as an individual's state of health, each of which may configure the others [20]. The built environment influences an individual's lifestyle and health behavior such as diet, physical activity, and active travel and configures the social environment, thereby influencing social interactions and support and hence an individual's mental health [21]. Therefore, the challenge for Chinese cities (as they evolve) will be to effectively integrate health factors into the urban planning of physical space in the ongoing phase of new urbanization, with an overall objective to enhance mental wealth of cities.

Studies have shown that housing conditions directly or indirectly affect people's mental health [22]. Previous research on the impact of housing conditions on mental health, primarily in the Western context, have considered the role of housing typology such as villa, duplex, apartment, and tenureship [23–26], floor-level [27–29], physical conditions [16,30–34], and access to residential facilities [35–38]. Specifically, Richman found that women living in villas had better mental health than women living in duplex and apartment houses [23]. Also, there are studies that found that residents living in high-rises have poor mental health mainly due to lack of security and social support [39]. Halpern divided 117 low-income families into experimental and reference groups. He reported that improving the housing conditions of the experimental group, such as increasing size of the modern kitchen and bathroom and providing constant hot water resulted in improvements in the mental health of participants of the experimental group, while the mental health of reference groups did not change significantly [40].

The housing environment may effect mental via several potential mechanisms.

(1) Amplifying adverse effects of social isolation. Social isolation and the associated lack of social relations may be a first cause of mental disorders [41]. Many studies have found that living in a high-rise leads to social isolation and can even affect a resident's social relationships and weaken people's mental health [27,42–44].

(2) Facilitating social contact and support. Social support may act as buffers against stress, thereby ameriolating adverse effects on mental health and wellbeing [45–48]. Through the exploration of the relationship between housing conditions and mental health of 279 housing area shortage residents, Smith found that in terms of the good or middle housing conditions, social support had a mediated effects between housing conditions and mental health, but when the housing conditions are not right, social support cannot alleviate the adverse effects of poor housing conditions on mental health [22].

(3) As a source of physiological and psycho-social stress. The housing micro-environment may directly act as source of stressors [49], and its relationship with mental wellbeing has long been recognized [50,51]. Studies have found associations between elevated levels of psychological distress and stress-related housing conditions including high residential density measured as persons/room [52–54] and related unwanted social interactions [55], physical quality of housing [31], and insecurity from crime [56].

(4) Perception and self-assessment has also been know produce effects on an individual's mental health [57], with low self-assessment being associated with mental health problems such as depression, suicide, irregular diet, and anxiety. High self-assessment has been known to improve mental health by offseting many adverse effects [58].

Prior research, mostly in Western contexts, has also shown that in addition to housing conditions themselves. The macro-level environment within the residential neighbourhood is a key determinant of health [59]. The characteristics of the community itself may also have an impact on individual health. Specifically, access to salutogenic community green space can provide residents with a healthy recreational environment and the opportunity to interact with others, both of which are good for mental health. From the perspective of the environment, good community green space, sanitation conditions, and security encourage more healthy activity behaviors such as exercise, walking, and active travel, which have the potential to improve a resident's physical and mental health. Hence, ignoring the effects of the macro-level community environment can lead to considerable confoundings [60]. Furthermore, the effects are complex, and community-level environment may interact with individual characteristics [61].

In 2015, the resident population urbanization rate in Shanghai was more than 89%, which is close to the level experienced in most developed countries. Housing conditions for the external migrant population continue to remain a major challenge [62,63]. However, there have thus far been no systematic study on the relationship between housing conditions, the macro-level neighborhood environment, and the mental health of migrants in high density Shanghai. To fill this research gap, we hypothesize that mental health of an individual is a function of micro- and macro-level environments. Direct effects include the effects of housing conditions such as the form of housing type and quality at micro-level directly effecting the trajectory of mental health while, macro-level neighborhood environment configure the level an individual's neighborhood satisfaction (sense of community) and accrued social capital, indirectly effecting mental health. We further hypothesize that housing conditions may have a direct and/or indirect effects upon mental health, which may be mediated by neighborhood satisfaction, and the mechanisms may be different in the cases of locals and migrants. The present empirical study uses questionnaire survey data derived from the outer area of Shanghai Jiading Nanxiang town. Structural equation modelling (SEM) is a relatively sophisticated (as compared to conventional regression analyses) methodology to model complex effects. We employed SEM to examine the causal pathways between housing conditions, neighborhood environment, and mental health. The conceptual model is illustrated in Figure 1. The objectives of the study are three-fold.

(1) To explore whether housing conditions have a direct effect on mental health;
(2) To examine the effects of housing upon mental health via indirect pathways through neighborhood satisfaction;
(3) To examine potential differences in the effects of housing conditions on mental health (strength and significance of effect estimates) of local and migrant population sub-groups.

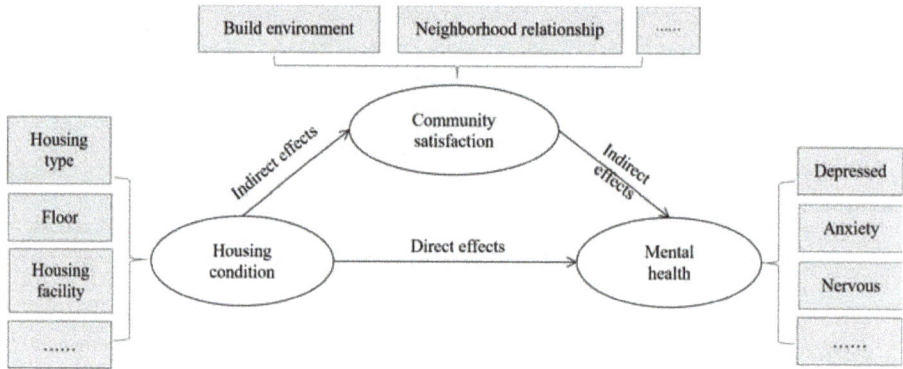

Figure 1. Concept model.

2. Data and Methods

2.1. Study Area and Data Sources

According to the sixth national census data, the city of Shanghai has an area of 6340.50 km^2 and a total population of 23.02 million, of which the migrant resident population is 8.98 million, amounting to 39%. Most of Shanghai's new migrants are spatially clustered in the peripheral districts of the city [64]. Therefore, the present study selected participants from these typical areas: Shanghai Jiading Nanxiang Town (Figure 2). A survey was conducted in 2015 employing stratified sampling (STR sampling) and proportional sampling (PPS sampling) for Shanghai local residents and migrants aged 15 or above and living in Nanxiang Town, including the village committee, neighborhood committee, and workplace (enterprise) as the study sampling sites (Figure 2), covering all six neighborhood committees, five village committees, and six major enterprises. In China, the village committee, neighborhood committee, and enterprise are the organizational clusters at the lowest level in rural, urban, and industrial areas. The total number of study participants was 320 (Table 1), with a series of valid questionnaires (293 in total) covering individual socio-demographics, lifestyle, residential environment, and mental health. After processing the missing values, the analytic sample comprised 251 participants, including 114 locals and 137 migrants, the proportion of migrants being 54.6%, which is close to the migrant's proportion (64.8%) in Nanxiang 2014 [65].

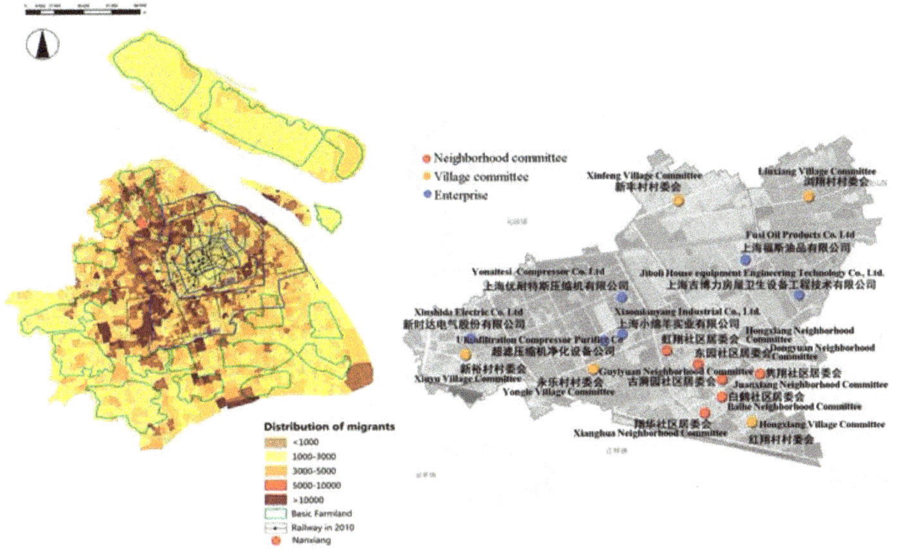

Figure 2. Questionnaire distribution map.

Table 1. Distribution of study participants.

Neighborhood Committee	Number	Village Committee	Number	Enterprise	Number
Guyiyuan	7	Xinyu	27	Giboli	17
Dongyuan	6	Yongle	48	Xinshida	28
Juanxiang	9	Hongxiang	16	Fuxiyoupin	21
Baihe	8	Xinfeng	8	Xiaomianyang	22
Xianghua	11	Liuxiang	23	Chaolv	22
Hongxiang	8	/	/	Younaitesi	39
Total of neighborhood committee questionnaires	49	Total of village committee questionnaires	122	Total of enterprise questionnaires	149
Total	320				

2.2. Statistical Method

As mentioned previously, we measured the direct effects of micro-level housing conditions and the indirect effect of the macro-level community environment captured via neighborhood satisfaction based upon mental health indicators. As neighborhood satisfaction is influenced by housing conditions and itself will affect mental health, it can efficiently help to understand the mediated effects. In this paper, the structural equation model (SEM) is used to verify the hypothesis. It is more suitable for dealing with multiple causes and multiple outcomes and involves unobserved latent variables that are indirectly inferred from multiple observed indicator variables [66]. The structural equation model is mainly composed of two parts: the measurement model and the structural model.

$$y = \Lambda_y \eta + \varepsilon \tag{1}$$

$$x = \Lambda_x \xi + \delta \tag{2}$$

$$\eta = B\eta + \Gamma\xi + \zeta \tag{3}$$

The measurement models (1) and (2) is about the relation between the index and the latent variable, and the structural model (3) imputes the relationship between the latent variables. In this study, the group SEM model was used to study the relationship between housing conditions and mental health of locals and migrants in Shanghai. Housing conditions, neighborhood satisfaction, and mental health were used as latent variables.

2.3. Measuring Housing Conditions

Participants' exposure to specific housing conditions were assessed through a series of questionnaires. Micro-level housing conditions were assessed through the following variables:

(1) Housing typology: This was assessed from participants' response to the question "Your current housing type is: (a) general building; (b) bungalow; (c) hut; (d) basement; (e) other." It can be seen from Table 2 that the locals are superior to the migrants regarding the type of housing. Compared with locals, more migrants live in cottage or sheds. Also, the probability of locals and migrants live in a general building is 0.906 and 0.643, respectively. Moreover, for both locals and migrants, the general building is the main type of housing, so to facilitate the model better fit, we divided housing type into "1" and "0" based on whether the housing type is general building or not.

(2) Residential building area in square meters was derived from the participants' response to the question "What is the size (in square meters) of your residential unit? If you live in a shared household, only estimate the living area of your family." Residential unit building area was employed as a continuous variable in our models.

(3) Access to basic residential facilities was assessed from a question related to seven residential facility typologies: "Does the house has separate facilities (namely, separate kitchen, separate washroom, shower facilities, access to liquefaction/pipe gas, air conditioning/heating equipment, home balcony, and elevator)," with 1 for a positive and 0 for negative response. Responses to these seven indicators were added as a dummy variable. The CFA test of the nine variables included in the housing condition has a CFI of 0.967 and greater than 0.95, which means the fitting result is good and the latent variables are set up reasonably. These nine indicators can describe the housing quality of locals and migrants.

Table 2. Housing type among the local and migrant categories.

Housing Type	Local	Migrant
general building	108	101
Bungalow	2	26
Hut	0	6
Basement	0	0
Other	7	24
Total	117	157

Neighborhood satisfaction was operationalized from participant responses to questionnaires related to satisfaction with macro-environmental aspects of housing, community facilities, and services. Many studies have shown that access to community facilities and services with residential environments, such as walkable landuses, salutogenic green space, etc., have a beneficial effect on resident's mental health, in particular for the elderly, women, and those living in low socioeconomic areas [27,36]. In the present study, satisfaction with community facilities and services was employed as a proxy for neighbourhood satisfactions. A set of seven questions were employed to measure satisfaction with various key community facilities: namely, community services, shopping and commercial facilities, community policing, sanitation, recreational facilities, community greening, and property management. Participant responses to these Likert scale questions ranged between 1–5 points (1 = dissatisfied, 2 = dissatisfied, 3 = so-so, 4 = satisfied, 5 = very satisfied). The CFI value of the scale is 0.995, indicating a good fit with the seven indicators acting as a reasonable proxy of neighborhood satisfaction for the locals and migrants.

2.4. Outcome: Mental Health

On the topic of mental health, the respondents were asked 12 questions about the extent of their current emotions based on the World Health Organization's mental health scale. From the 12 indicators, we selected five indicators for this study to measure mental health. The Mental Health Scale is a general instrument for assessing the psychological problems of ordinary people [28], which has been widely used in large sample mental health surveys such as the National Health Survey and the World Health Organization's World Mental Health Survey (WMH) [29]. The scale contains questions on the frequency of nervousness, despair, anxiety, or upset over the last 30 days. According to the mental health scale and the best fit of our actual data, we selected five indicators as the main indicators of mental health in the present study. These were anxiety and insomnia, nervousness, cannot overcome difficulties, unhappiness and depression, and loss of confidence. Response to each question was based on the frequency of the problem and reported as never, rarely, sometimes, and often, each given 4–1 points. The higher the score, the better the mental health. The reported reliability of the scale was 0.845, and the CFI value of the CFA test was 0.971, indicating a good fit.

3. Analysis Results

3.1. Sample Characteristics

Table 3 summarizes the residential environment indicators among the 114 locals and the 137 migrants in Nanxiang Town. A non-parametric Kruskal-Wallis test was employed to examine the level of statistical significance of the differences of means across categories. The results of Table 3 indicate that there are significant differences between migrants and locals across indicators of housing conditions, neighborhood satisfaction, mental health, and demographic indicators. Most of the average or quantitative indicators of migrants are lower than those of Shanghai locals. Compared with locals, migrants are exposed to a poorer residential environment. In fact, these poorer housing conditions are reflected in poor housing conditions and in living areas, residential discrimination, and other aspects [30].

Table 3. Descriptive characteristics of the participants' residential environment.

Variables	Local (*n* = 114)	Migrant (*n* = 137)	*p*-Value of Difference
Housing conditions			
Commercial residential building; *N* (%)	108.00 (94.7%)	101.00 (73.7%)	***
housing area; Mean (SD)	53.70 (30.77)	47.10 (24.14)	*
Facilities; *N* (%)			
Separate kitchen	106.00 (93.0%)	93.00 (67.9%)	***
Separate toilet	105.00 (92.1%)	87.00 (63.5%)	***
Shower	105.00 (92.1%)	96.00 (70.1%)	***
Gas	104.00 (91.2%)	82.00 (59.9%)	***
Air condition	103.00 (90.4%)	95.00 (69.3%)	***
Balcony	96.00 (84.2%)	69.00 (50.4%)	***
Elevator	39.00 (34.2%)	37.00 (27.0%)	
Neighborhood satisfaction; Mean (SD)			
Community services	3.67 (0.88)	3.20 (0.92)	***
Commercial facilities	4.35 (0.96)	4.15 (1.15)	
Community policing	3.60 (0.90)	3.50 (0.92)	
Sanitary condition	3.53 (0.92)	3.34 (0.93)	
Recreational facilities	3.32 (0.99)	3.13 (0.99)	
Green space	3.50 (1.01)	3.31 (0.95)	
Property management	4.36 (0.96)	4.10 (0.99)	*

Table 3. *Cont.*

Variables	Local (*n* = 114)	Migrant (*n* = 137)	*p*-Value of Difference
Mental health; Mean (SD)			
Insomnia because of anxiety	2.82 (0.90)	2.71 (0.88)	
Always feel nervous	1.97 (1.19)	1.76 (1.09)	
Cannot overcome difficulties	2.68 (0.80)	2.61 (0.79)	
Be unhappy and depressed	2.83 (0.89)	2.66 (0.79)	*
Lose confidence	3.01 (0.92)	3.07 (0.82)	
Age; Mean (SD)	40.21 (13.14)	31.83 (7.22)	***
Gender; *N* (%) (female as reference)	58.00 (50.9%)	73.00 (53.3%)	
Income (ln); Mean (SD)	1.50 (0.67)	1.75 (0.51)	**

* $p < 0.05$; ** $p < 0.01$; *** $p < 0.001$ two-tailed *t*-tests chi-square test.

Specifically, the differences between the migrants and locals are mainly reflected in the housing conditions. Housing tenureship, quality, and access to facilities were poorer in the migrant category as compared to the locals. Locals mostly resided in self-owned housing (93.7%), while migrants mainly lived in rented accommodation (78.8%). Results indicated that 59.1% of migrants were sole tenants, 19.7% were joint tenants, and the rent of 61.6% tenant was less than 1000 yuan. Morover, locals are significantly more likely to live in commercial housing compared with migrants. The choice of housing for migrants were more biased, with low rent corresponding to increasing commuting distance to place of work. Furthermore, because of the lower socioeconomic status and prominent liquidity characteristics of migrants, their residential environment had a temporary characteristic, resulting in significantly poorer housing quality as compared to non-migrant residents [31]. Also, there was a significant difference in neighborhood satisfaction scores rating community service, property management, and community facilities satisfaction, with migrants scoring significantly lower than Shanghai locals. With respect to mental health, the results of the mental health comparison between residents and foreign population in Nanxiang are similar to those of domestic scholars. There are also some differences in mental health between migrants and locals [32–36]. "Unhappy and depressed" scores for migrants (at the $\alpha = 0.05$ level) was significantly lower than those of locals, implying there is a higher prevalence of unhappiness and depression among migrants. The mean age of the population also significantly differed between locals and migrants, with the latter's average age being eight years younger than that of the locals, with participants between 16 to 60 years old accounting for 96.8% of the total population sample. As expected, there were significant differences in income between the local and migrant populations.

3.2. Analysis Results

Model fit statistics indicate a good fit of the a priori SEM model, with a Chi square test of model fit at 844.50 on 486 degrees of freedom and RMSEA of 0.084. The model fitting index was close to one—CFI = 0.880 and TLI = 0.875. The structural equations model illustrated in Figures 3 and 4 examines the impact of housing conditions and neighborhood satisfaction on mental health among Shanghai locals and migrants, respectively. The study reported differences in the mechanism by which housing conditions affect mental health among locals and migrants.

For Shanghai locals (Figure 3), the path from housing conditions to neighborhood satisfaction was not significant. It was found that the direct effect of housing condition on local participants' mental health was not significant, at 95% confidence level ($p = 0.055$), and borderline significance ($p < 0.10$) indicated the mild protective effects of better housing conditions upon mental health. As per expectations, the path coefficient of neighborhood satisfaction upon the mental health was 0.423 ($p = 0.000$), pointing to accrued beneficial effects upon mental health. Moreover, the older the person is, the better his mental health will be, while age had no significant effect on neighborhood satisfaction. Income was observed to directly effect locals' neighborhood satisfaction but did not influence mental health.

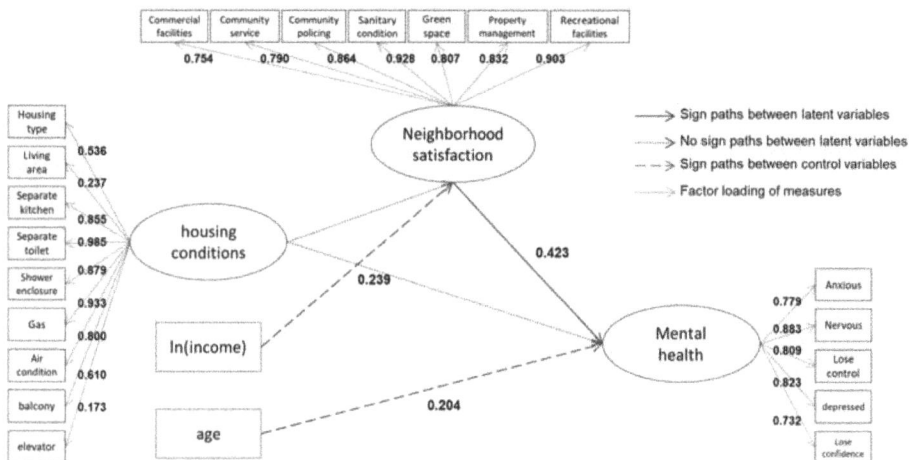

Figure 3. Structural equation model linking housing characteristics to mental health among Shanghai locals.

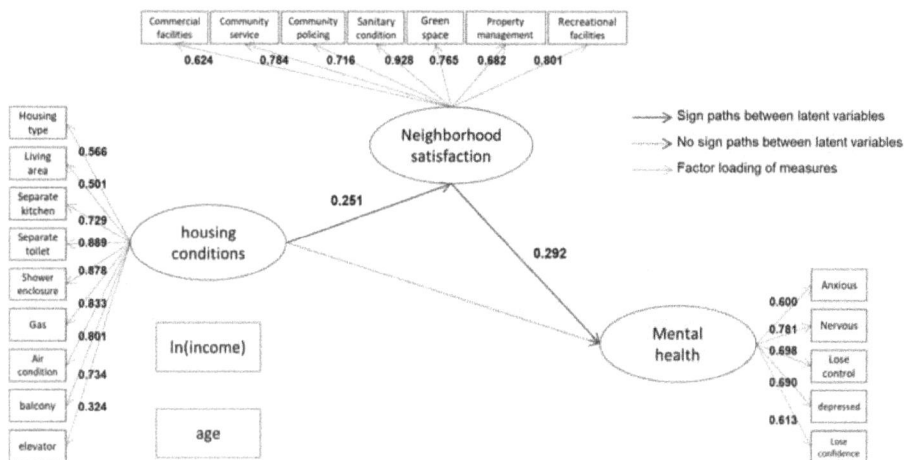

Figure 4. Structural equation model linking housing characteristics to mental health among Shanghai migrants.

On the contrary, in the case of migrants (Figure 4), housing conditions per se had no significant direct effect upon mental health ($p = 0.187$), the effects were mediated via neighborhood satisfaction. Housing conditions were linked with neighborhood satisfaction with a path coefficient of 0.251 ($p = 0.013$), and the direct path coefficient from neighborhood satisfaction to mental health was 0.292 ($p = 0.006$), which may imply that although the housing conditions of migrants do not directly affect mental health, improving the housing conditions of migrants constitutes a better perception of neighborhood, and increments in neighborhood satisfaction have a significant positive impact on mental health. The effects of age and income on community satisfaction and mental health of migrants remained insignificant.

The results are summarized in Table 4. In a nutshell, in the case of migrant participants. housing conditions had an indirect effect upon mental health, the effect being mediated by neighborhood satisfaction. For local participants, housing conditions were not linked with neighborhood satisfaction, while income had indirect effect on mental helath via neighborhood satisfaction.

Table 4. Results of impact analysis.

Variable	Neighborhood Satisfaction	Mental Health					
		Direct Effect		Indirect Effect		Total Effect	
Local							
Housing condition	0.54	0.60		0.20		0.80	**
Age	−0.00	0.01	*	−0.00		0.01	*
Income(ln)	0.28	*	−0.10		0.11	*	0.00	
Migrants							
Housing condition	0.48	*	−0.23		0.11		−0.11	
Age	0.01	0.01		0.00		0.01	*
Income(ln)	0.01	−0.11		0.00		−0.11	

* $p < 0.05$; ** $p < 0.01$; N = 251.

4. Discussion

In the present study, we examined the effects of housing conditions upon mental health among participants from Nanxiang Town in Shanghai through a structural equation modelling approach. The application of SEM enabled isolation of both direct and indirect pathways along which housing condition influences on people's mental health. Specifically, our results showed that housing conditions have indirect effects on migrant's mental health via neighborhood satisfaction, while for local participants, housing conditions posed a direct effect on mental health.

Our study provides substantial evidence that housing conditions in China act along two distinct causal pathways to produce effects upon individual's mental health [22]. Neighborhood satisfaction was reported to mediate the effect of housing conditions on the mental health of Shanghai migrants. The results compliment prior evidence that showed that neighborhood effect does matter for social integration and cohesion issues in Shanghai [67,68]. They point to the fact that a community with greater satisfaction can provide a sense of belonging to its inhabitants, and enhanced social support can mitigate the psychological pressure derived from poor housing conditions [12,39].

The principle finding that two specific mechanisms exist in the case of local and migrant populations, respectively, where housing conditions effect mental health is of significant relevance. In principle, this points to the underlying causal pathways. Local residents are associated with residential stability and, as our data shows, have access to better quality housing and neighborhood environments. Housing conditions did not directly or indirectly produce significant effects on mental health, while neighborhood satisfaction directly influenced mental health. The relationship between income and mental health was also mediated via neighborhood satisfaction. However, in the case of migrants, housing conditions did not directly affect mental health but had significant indirect effects mediated via neighborhood satisfaction. Migrants are more likely to self-select housing; the choice is dependent mostly on economic considerations of affordability such as residential rent instead of housing quality per se [69]. The detected relationship between housing conditions and neighborhood satisfaction in migrants is indicative of the socio-spatial clustering of poor migrant housing in deprived neighborhoods as experienced in Shanghai. The null direct effects of housing conditions may be explained on the basis of lack of affordability, a sort of psychological saturation wherein there is no expectation among the migrants of better quality housing. It may also be related to the actual time spent in residences. Often, migrant workers do multiple jobs, spending longer hours outside for sustenance with significantly less time spent within the residence. The observed indirect effects of housing conditions mediated by neighborhood satisfaction and the direct effects of the later upon mental health may be a result of a number of underlying factors. First, migrants are inherently associated with broken social network [5] when they migrate to cities, since they generally do not bring relatives with them, their children still stay in their hometown [28], and hence migrants have weak social tie support. Second, poorer economic conditions imply they are segregated to living in deprived neighborhoods with a reduced sense of community and safety (higher fear of crime). Third, unstable job contracts mean lower residential stability and higher rate of migration between neighborhoods with generally similar housing conditions. In light of these factors, a healthy community environment—one that promotes sense of community and social support and is associated with enhanced perception of territoriality, defensible space, and safety—is of critical importance in mental wellbeing. In contrast, housing conditions, community environment, and community services are the primary consideration in local people's choices regarding accommodations. We acknowledge the cross-sectional design and small scale as limitations of the present study. Largescale studies involving more townships with longitudinal design are needed to further validate these findings. Such largescale studies with significant heterogeneity in population characteristics must also adjust for differences in lifestyles arising from inherent cultural affinities.

5. Conclusions

Mental wealth constitutes an indispensable resource of any sustainable global city. In a rapidly urbanizing city such as Shanghai, the mental health of the migrant workforce is one of the primary considerations in the economics of developmental pressures versus socio-spatial and health disparities. Our findings have significant policy implications, especially at the level of community health planning and management. As an important upstream-level community intervention to minimize health disparities, it is important to optimize the density and mix of local-to-migrant housing to promote a sense of community and enhance people's social capital and wellbeing [69]. Provisioning access to key community services is also significant; for instance, with the evidence confirming the environment injustice and inefficient access to public parks in Shanghai [70–73], the local government should aim to improve the quality of green services to provide a livable and inclusive environment promoting mental health of the migrant workers and hence enhancing the overall mental capital of the city. At an institutional level, policy makers have a role to further refine policies related to *hukou* registration with a view to improve access to basic services (reduce their disparities) that adversely affect migrant workers' mental health.

As Chinese cities continue to rapidly urbanize, planning sustainable housing provisions for migrant workers will ensure social sustainability via reducing disparity and accrue significant economic benefits through reductions in chronic disease and mental illnesses among vulnerable migrants and related costs of treatment and care in the long run.

Acknowledgments: This work was supported by the National Natural Science Foundation of China under Grant (No. 41501170), the Fundamental Research Funds for the Central Universities, and Hong Kong University's URC Research Assistant Professorship grant.

Author Contributions: Chinmoy Sarkar conceived and designed the experiments; Huizhi Geng and Yi Lu contributed to survey design and its implement; Siyu Miao and Yang Xiao analyzed the data; Siyu Miao and Yang Xiao wrote the paper.

Conflicts of Interest: The authors declare no conflict of interest.

References

1. National Bureau of Statistics of China. *China City Statistical Yearbook 2016*; China Statistics Press: Beijing, China, 2017.
2. Chatterjee, D.K. *United Nations Development Program (UNDP)*; Springer: Dordrecht, The Netherlands, 2011; pp. 548–549.
3. Maridal, J.H. A Worldwide Measure of Societal Quality of Life. *Soc. Indic. Res.* **2016**, *134*, 1–38. [CrossRef]
4. Wirth, L. Urbanism as a Way of Life. *Am. J. Soc.* **1938**, *44*, 1–24. [CrossRef]
5. Van, O.J.; Hanssen, M.; Bak, M.; Bijl, R.V.; Vollebergh, W. Do urbanicity and familial liability coparticipate in causing psychosis? *Am. J. Psychiatry* **2003**, *160*, 477–482. [CrossRef] [PubMed]

6. Sundquist, K.; Frank, G.; Sundquist, J. Urbanisation and incidence of psychosis and depression: Follow-up study of 4.4 million women and men in Sweden. *Br. J. Psychiatry J. Ment. Sci.* **2004**, *184*, 293–298. [CrossRef]
7. Beddington, J.; Cooper, C.L.; Field, J. The mental wealth of nations. *Nature* **2008**, *455*, 1057–1060. [CrossRef] [PubMed]
8. Bairoch, P. *Cities and Economic Development: From the Dawn of History to the Present*; University of Chicago Press: Chicago, IL, USA, 1991.
9. Okulicz-Kozaryn, A. *Happiness and Place: Why Life Is Better Outside of the City*; Palgrave Macmillan U.S.: New York, NY, USA, 2015.
10. Yang, G.; Wang, Y.; Zeng, Y.; Gao, G.F.; Liang, X.; Zhou, M.; Wan, X.; Yu, S.; Jiang, Y.; Naghavi, M.; et al. Rapid health transition in China, 1990–2010: Findings from the Global Burden of Disease Study 2010. *Lancet* **2013**, *381*, 1987–2015. [CrossRef]
11. Gu, L.; Xie, J.; Long, J.; Chen, Q.; Chen, Q.; Pan, R.; Yan, Y.; Wu, G.; Liang, B.; Tan, J.; et al. Epidemiology of major depressive disorder in mainland china: A systematic review. *PLoS ONE* **2013**, *8*, e65356. [CrossRef] [PubMed]
12. Zhong, B.L.; Liu, T.B.; Chan, S.S.M.; Jin, D.; Hu, C.Y.; Dai, J.; Chiu, H.F.K. Prevalence and correlates of major depressive disorder among rural-to-urban migrant workers in Shenzhen, China. *J. Affect. Disord.* **2015**, *183*, 1–9. [CrossRef] [PubMed]
13. McKenzie, K. Urbanization, social capital and mental health. *Glob. Soc. Policy* **2008**, *8*, 359–377. [CrossRef]
14. Bhugra, D. Migration and mental health. *Acta Psychiatr. Scand.* **2004**, *109*, 243–258. [CrossRef] [PubMed]
15. Nielson, I.; Nyland, C.; Smyth, R.; Zhang, M.; Zhu, C.J. Which rural migrants receive social insurance in Chinese cities? Evidence from Jiangsu survey data. *Glob. Soc. Policy* **2005**, *5*, 353–381. [CrossRef]
16. Evans, G.W.; Kantrowitz, E.; Eshelman, P. Housing Quality and Psychological Well-Being among the Elderly Population. *J. Gerontol.* **2002**, *57*, 381–383. [CrossRef]
17. Allen, J.; Balfour, R.; Bell, R.; Marmot, M. Social determinants of mental health. *Int. Rev. Psychiatry* **2014**, *4*, 392–407. [CrossRef] [PubMed]
18. Evans, G.W. The built environment and mental health. *J. Urban Health* **2003**, *80*, 536–555. [CrossRef] [PubMed]
19. Berkman, L.F.; Glass, T.; Brissette, I.; Seeman, T.E. From social integration to health: Durkheim in the new millennium. *Soc. Sci. Med.* **2000**, *51*, 843–857. [CrossRef]
20. Vlahov, D.; Galea, S. Urbanization, urbanicity, and health. *J. Urban Health* **2002**, *79*, S1–S12. [CrossRef] [PubMed]
21. Sarkar, C.; Webster, C.; Gallacher, J. *Healthy Cities: Public Health through Urban Planning*; Edward Elgar Pub.: Cheltenham, UK, 2014; pp. 319–324.
22. Smith, C.A.; Smith, C.J.; Kearns, R.A.; Abbott, M.W. Housing stressors, social support and psychological distress. *Soc. Sci. Med.* **1993**, *37*, 603–612. [CrossRef]
23. Richman, N. Behaviour problems in pre-school children: Family and social factors. *Br. J. Psychiatry* **1977**, *131*, 523–527. [CrossRef] [PubMed]
24. Ellaway, A.; Macintyre, S. Does housing tenure predict health in the UK because it exposes people to different levels of housing related hazards in the home or its surroundings. *Health Place* **1998**, *4*, 141–150. [CrossRef]
25. Yanos, D.P.T.; Felton, B.J.; Tsemberis, S.; Frye, V.A. Exploring the role of housing type, neighborhood characteristics, and lifestyle factors in the community integration of formerly homeless persons diagnosed with mental illness. *J. Ment. Health* **2007**, *16*, 703–717. [CrossRef]
26. Baker, E.; Bentley, R.; Mason, K. The mental health effects of housing tenure: Causal or compositional? *Urban Stud.* **2013**, *50*, 426–442. [CrossRef]
27. Gillis, A.R. High-rise housing and psychological strain. *J. Health Soc. Behav.* **1977**, *18*, 418–431. [CrossRef] [PubMed]
28. Kearns, A.; Whitley, E.; Mason, P.; Bond, L. 'Living the high life'? Residential, social and psychosocial outcomes for high-rise occupants in a deprived context. *Hous. Stud.* **2002**, *27*, 97–126. [CrossRef]
29. Panczak, R.; Galobardes, B.; Spoerri, A.; Zwahlen, M.; Egger, M. High life in the sky? Mortality by floor of residence in Switzerland. *Eur. J. Epidemiol.* **2013**, *28*, 453–462. [CrossRef] [PubMed]
30. Weich, S.; Blanchard, M.; Prince, M.; Burton, E.; Erens, B.; Erens, B.; Sproston, K. Mental health and the built environment: Cross-sectional survey of individual and contextual risk factors for depression. *Br. J. Psychiatry* **2002**, *180*, 428–433. [CrossRef] [PubMed]

31. Galea, S.; Ahern, J.; Rudenstine, S.; Wallace, Z.; Vlahov, D. Urban built environment and depression: A multilevel analysis. *J. Epidemiol. Community Health* **2005**, *59*, 822–827. [CrossRef] [PubMed]
32. Guite, H.F.; Clark, C.; Ackrill, G. The impact of the physical and urban environment on mental well-being. *Public Health* **2006**, *120*, 1117–1126. [CrossRef] [PubMed]
33. Araya, R.; Montgomery, A.; Rojas, G.; Fritsch, R.; Solis, J.; Signorelli, A.; Lewis, G. Common mental disorders and the built environment in Santiago, Chile. *Br. J. Psychiatry* **2007**, *190*, 394–401. [CrossRef] [PubMed]
34. Rollings, K.A.; Wells, N.M.; Evans, G.W.; Bednarz, A.; Yang, Y. Housing and neighborhood physical quality: Children's mental health and motivation. *J. Environ. Psychol.* **2017**, *50*, 17–23. [CrossRef]
35. Saarloos, D.; Alfonso, H.; Giles-Corti, B.; Middleton, N.; Almeida, O.P. The Built Environment and Depression in Later Life: The Health In Men Study. *Am. J. Geriatr. Psychiatry* **2011**, *19*, 461–470. [CrossRef] [PubMed]
36. Sarkar, C.; Gallacher, J.; Webster, C. Urban built environment configuration and psychological distress in older men: Results from the Caerphilly study. *BMC Public Health* **2013**, *13*, 695. [CrossRef] [PubMed]
37. Gariepy, G.; Thombs, B.D.; Kestens, Y.; Kaufman, J.S.; Blair, A.; Schmitz, N. The Neighbourhood Built Environment and Trajectories of Depression Symptom Episodes in Adults: A Latent Class Growth Analysis. *PLoS ONE* **2015**, *10*, e0133603. [CrossRef] [PubMed]
38. Wu, Y.T.; Prina, A.M.; Jones, A.; Matthews, F.E.; Brayne, C. The Built Environment and Cognitive Disorders: Results from the Cognitive Function and Ageing Study II. *Am. J. Prev. Med.* **2017**, *53*, 25–32. [CrossRef] [PubMed]
39. McCarthy, D.; Saegert, S. Residential density, social overload, and social withdrawal. *Hum. Ecol.* **1978**, *6*, 253–272. [CrossRef]
40. Halpern, D. Mental Health and the Built Environment. *Psychiatr. Serv.* **1997**, *48*, 721.
41. Umberson, D.; Montez, J.K. Social relationships and health a flashpoint for health policy. *J. Health Soc. Behav.* **2010**, *51* (Suppl. 1), S54–S66. [CrossRef] [PubMed]
42. McCarthy, P.; Byrne, D.; Harrison, S.; Keithley, J. Housing type, housing location and mental health. *Soc. Psychiatry Psychiatr. Epidemiol.* **1985**, *20*, 125–130. [CrossRef]
43. Husaini, B.A.; Moore, S.T.; Castor, R.S. Social and Psychological Well-Being of Black Elderly Living in High-Rises for the Elderly. *J. Gerontol. Soc. Work* **1991**, *16*, 57–78. [CrossRef]
44. Mita, T.J. Mental health and high flats. *J. Chronic Dis.* **1982**, *35*, 495. [CrossRef]
45. Turner, R.J.; Frankel, B.G.; Levin, D.M. Social support: Conceptualization, measurement, and implications for mental health. *Res. Community Ment. Health* **1983**, *3*, 67–111. [CrossRef]
46. Cohen, S.; Wills, T.A. Stress, social support, and the buffering hypothesis. *Psychol. Bull.* **1985**, *98*, 310–357. [CrossRef] [PubMed]
47. Dalgard, O.S.; Bjørk, S.; Tambs, K. Social support, negative life events and mental health. *Br. J. Psychiatry J. Ment. Sci.* **1995**, *166*, 29–34. [CrossRef]
48. Kawachi, I.; Berkman, L.F. Social ties and mental health. *J. Urban Health Bull. N. Y. Acad. Med.* **2001**, *78*, 458–466. [CrossRef] [PubMed]
49. Robinson, E.; Adam, R.; Australian Family Relationships Clearinghouse; Australian Institute of Family Studies. *Housing Stress and the Mental Health and Wellbeing of Families*; Australian Institute of Family Studies: Melbourne, Australia, 2008.
50. Aneshensel, C.S.; Rutter, C.M.; Lachenbruch, P.A. Social structure, stress, and mental health: Competing conceptual and analytic models. *Am. Sociol. Rev.* **1991**, *56*, 166–178. [CrossRef]
51. Slavich, G.M.; Irwin, M.R. From Stress to Inflammation and Major Depressive Disorder: A Social Signal Transduction Theory of Depression. *Psychol. Bull.* **2014**, *140*, 774–815. [CrossRef] [PubMed]
52. Lepore, S.J.; Evans, G.W.; Schneider, M.L. Dynamic role of social support in the link between chronic stress and psychological distress. *J. Personal. Soc. Psychol.* **1991**, *61*, 899–909. [CrossRef]
53. Evans, G.W.; Lepore, S.J. Household crowding and social support: A quasiexperimental analysis. *J. Personal. Soc. Psychol.* **1993**, *65*, 308–316. [CrossRef]
54. Fuller, T.D.; Edwards, J.N.; Vorakitphokatorn, S.; Sermsri, S. Chronic stress and psychological well-being: Evidence from Thailand on household crowding. *Soc. Sci. Med.* **1996**, *42*, 265–280. [CrossRef]
55. Saegert, S. Environments and children's mental health: Residential density and low income children. In *Handbook of Psychology and Health*; Baum, A., Singer, J.E., Eds.; Lawrence Erlbaum: Hillsdale, NJ, USA, 1982.
56. Cohen, D.A.; Mason, C.; Bedimo, A.; Scribner, R.; Basolo, V.; Farley, T.A. Neighborhood physical conditions and health. *Am. J. Public Health* **2003**, *93*, 467–471. [CrossRef] [PubMed]

57. Kearns, A.; Ellaway, A.; Macintyre, S.; Hiscock, R. The THAW Report: Findings of a study of transport, housing, and well being in the West of Scotland. *ChemBioChem* **2000**, *16*, 2329–2336.

58. Mann, M.M.; Hosman, C.M.H.; Schaalma, H.P.; De Vries, N.K. Self-esteem in a broad-spectrum approach for mental health promotion. *Health Educ. Res.* **2004**, *19*, 357–372. [CrossRef] [PubMed]

59. Krieger, J.; Higgins, D.L. Housing and health: Time again for public health action. *Am. J. Public Health* **2002**, *92*, 758–768. [CrossRef] [PubMed]

60. Pickett, K.; Pearl, M. Multi-level analyses of neighborhood socioeconomic context and health outcomes: A critical review. *J. Epidemiol. Community Health* **2001**, *55*, 111–122. [CrossRef] [PubMed]

61. Wen, M.; Fan, J.; Jin, L.; Wang, G. Neighborhood effects on health among migrants and natives in Shanghai, China. *Health Place* **2010**, *16*, 452–460. [CrossRef] [PubMed]

62. Wu, W. Migrant housing in urban China-Choices and constraints. *Urban Aff. Rev.* **2002**, *38*, 90–119. [CrossRef]

63. Wu, W. Sources of Migrant Housing Disadvantage in Urban China. *Environ. Plan. A* **2004**, *36*, 1285–1304. [CrossRef]

64. Xiao, Y.; Chen, S.; Wang, X.; Huang, J. Study on segregation of residential space of new migrants in Shanghai from the perspective of global city. *City Plan. Rev.* **2016**, *40*, 25–33.

65. Liu, Y. *Research on the Dwelling Characteristics of the Floating Population in Shanghai Suburbs and the Influence Factors of Changes in the Housing Conditions*; Tongji University: Shanghai, China, 2016.

66. Hancock, G.R. Fortune cookies, measurement error, and experimental design. *J. Mod. Appl. Stat. Methods* **2003**. [CrossRef]

67. Wang, Z.; Zhang, F.; Wu, F. Intergroup neighbouring in urban China: Implications for the social integration of migrants. *Urban Stud.* **2016**, *53*, 651–668. [CrossRef]

68. Wang, Z.; Zhang, F.; Wu, F. Social Trust between Rural Migrants and Urban Locals in China—Exploring the Effects of Residential Diversity and Neighbourhood Deprivation. *Popul. Space Place* **2017**, *23*. [CrossRef]

69. Liu, Y.; Zhang, F.; Wu, F.; Liu, Y.; Li, Z. The subjective wellbeing of migrants in Guangzhou, China: The impacts of the social and physical environment. *Cities* **2017**, *60*, 333–342. [CrossRef]

70. Wang, Z.; Zhang, F.; Wu, F. Affective neighbourly relations between migrant and local residents in Shanghai. *Urban Geogr.* **2016**. [CrossRef]

71. Xiao, Y.; Wang, Z.; Li, Z.; Tang, Z. An assessment of urban park access in Shanghai—Implications for the social equity in urban China. *Landsc. Urban Plan.* **2017**, *157*, 383–393. [CrossRef]

72. Xiao, Y.; Lu, Y.; Guo, Y.; Yuan, Y. Estimating the willingness to pay for green space services in Shanghai: Implications for social equity in urban China. *Urban For. Urban Green.* **2017**, *26*, 95–103. [CrossRef]

73. Xiao, Y.; Li, Z.; Webster, C. Estimating the mediating effect of privately-supplied green space on the relationship between urban public green space and property value: Evidence from Shanghai, China. *Land Use Policy* **2016**, *54*, 439–447. [CrossRef]

International Journal of
*Environmental Research
and Public Health*

MDPI

Article

Does Walkability Contribute to Geographic Variation in Psychosocial Distress? A Spatial Analysis of 91,142 Members of the 45 and Up Study in Sydney, Australia

Darren J. Mayne [1,2,3,4,*] [iD], Geoffrey G. Morgan [1,5], Bin B. Jalaludin [6,7] and Adrian E. Bauman [1]

1   Sydney School of Public Health, The University of Sydney, Sydney, NSW 2006, Australia;
    geoffrey.morgan@sydney.edu.au (G.G.M.); adrian.bauman@sydney.edu.au (A.E.B.)
2   Public Health Unit, Illawarra Shoalhaven Local Health District, Wollongong, NSW 2502, Australia
3   School of Medicine, University of Wollongong, Wollongong, NSW 2522, Australia
4   Illawarra Health and Medical Research Institute, University of Wollongong,
    Wollongong, NSW 2522, Australia
5   University Centre for Rural Health—North Coast, The University of Sydney, Sydney, NSW 2006, Australia
6   Ingham Institute, University of New South Wales, Sydney, NSW 2052, Australia; b.jalaludin@unsw.edu.au
7   Epidemiology, Healthy People and Places Unit, Population Health,
    South Western Sydney Local Health District, Liverpool, NSW 1871, Australia
*   Correspondence: dmay8519@uni.sydney.edu.au; Tel.: +61-2-4221-6733

Received: 27 December 2017; Accepted: 1 February 2018; Published: 6 February 2018

Abstract: Walkability describes the capacity of the built environment to promote walking, and has been proposed as a potential focus for community-level mental health planning. We evaluated this possibility by examining the contribution of area-level walkability to variation in psychosocial distress in a population cohort at spatial scales comparable to those used for regional planning in Sydney, Australia. Data on psychosocial distress were analysed for 91,142 respondents to the 45 and Up Study baseline survey between January 2006 and April 2009. We fit conditional auto regression models at the postal area level to obtain smoothed "disease maps" for psychosocial distress, and assess its association with area-level walkability after adjusting for individual- and area-level factors. Prevalence of psychosocial distress was 7.8%; similar for low (7.9%), low-medium (7.9%), medium-high (8.0%), and high (7.4%) walkability areas; and decreased with reducing postal area socioeconomic disadvantage: 12.2% (most), 9.3%, 7.5%, 5.9%, and 4.7% (least). Unadjusted disease maps indicated strong geographic clustering of psychosocial distress with 99.0% of excess prevalence due to unobserved and spatially structured factors, which was reduced to 55.3% in fully adjusted maps. Spatial and unstructured variance decreased by 97.3% and 39.8% after adjusting for individual-level factors, and another 2.3% and 4.2% with the inclusions of area-level factors. Excess prevalence of psychosocial distress in postal areas was attenuated in adjusted models but remained spatially structured. Postal area prevalence of high psychosocial distress is geographically clustered in Sydney, but is unrelated to postal area walkability. Area-level socioeconomic disadvantage makes a small contribution to this spatial structure; however, community-level mental health planning will likely deliver greatest benefits by focusing on individual-level contributors to disease burden and inequality associated with psychosocial distress.

Keywords: disease mapping; geographic variation; psychosocial distress; spatial analysis; walkability

1. Introduction

Mental illness is a leading cause of disability worldwide [1] accounting for 19% of total years lived with disability (YLD) and 7% of disability-adjusted life years (DALY) [2,3] of which 53% is due to depressive and anxiety disorders [4]. Just under one-half (45.5%) of the Australian adult population report having ever experienced a mental disorder in their lifetime and one-fifth (20%) in the previous 12 months [5]. In 2012, the World Health Organization challenged its member states to reduce their disability burdens due to mental illness through coordinated action between health and social sectors [6]. This was followed in 2013 by a comprehensive action plan that emphasised addressing the many determinants of mental illness, including environmental factors that contribute to individual and population-level vulnerabilities [7].

Walkability describes the capacity of the built environment to facilitate walking for various purposes, including transportation, health and leisure [8]. A small but growing literature has emerged over the last decade examining associations between walkability and mental health [9–12], leading some commentators to recommend walkability as a potential focus for community-level mental health planning and programming [13]. The current evidence base is insufficiently developed to identify a pathway by which walkability may influence mental health; however, two possibilities have been suggested. The first hypothesises that walkable environments help to promote positive affect by increasing participation in moderate-intensity physical activity, such as walking [9]. This is consistent with review findings that participation in regular physical activity protects against the onset of depression and anxiety in healthy populations, and reduces the severity of symptoms in clinical populations [14–16]; possibly by modulating melatonin production, adenosine metabolism, and circadian rhythms, or activating brain centres that help reduce negative affect [14]. The second hypothesises that walkable environments may enhance the social capital of neighbourhoods by providing unstructured opportunities for social interactions between individuals [17] that promote trust, and enhance feelings of familiarity, certainty, resilience, and reciprocity [9,17–19]. Social capital is understood to buffer individuals against depression and anxiety by reducing daily pressures and promoting health-enhancing behaviours [20]. However, despite their plausibility, neither hypothesis is currently supported by evidence from an appropriate causal evaluation.

Walkability is typically derived as an objective index within a geographical information system [21] using spatial data on residential dwelling density, street network connectivity, land use mix, and—when available—retail destinations, density or floor space [8,22,23]. Indexes originating out of the North American Neighborhood Quality of Life Study (NQLS) [22] and Australian Physical Activity in Localities and Community Environments (PLACE) Study [8] projects have contributed to an extensive evidence base within the transportation, planning, and public health literatures linking the walkability of built environments to improvements in health behaviours and outcomes [24–29]. Much of this evidence comes from individual-level studies of participants and the micro (personal) and meso (neighbourhood) environments in which they live [21,30]. However, there is increasing interest in meso (area) environment walkability, its contributions to the distribution of health within populations, and how it may be used to inform population health programming at larger regional scales [8,23,30–32].

Psychosocial or psychological distress describes anxious or depressed mood in the absence of a specific psychiatric diagnosis [33] and is commonly used to monitor mental health status in populations using representative surveys [34], such as the United States (US) Behavioral Risk Factor Surveillance System [35] and Australian Health Survey [36]. Environmental influences on mental health have received considerable attention in the research literature (see [19] for reviews); however, only a small number of studies have directly addressed relations between walkability and mental health outcomes [19], and none at the spatial scales typically used for population health planning and intervention. Between-group analyses of outcomes such as psychosocial distress can identify population sub-groups at increased risk of adverse mental health outcomes but provide limited information on the geography of these risks. In contrast, spatial analyses may be used to identify areas at increased risk of adverse outcomes or spatially structured influences on health by focusing on geographic variation in excess of that due to known demographic, social, economic, and health factors [30,37,38].

Spatial analyses of health outcomes and behaviours are increasingly common in the epidemiological literature as statistical methods and geographically-referenced administrative, surveillance and research data become more accessible [39]. Spatial analyses are especially informative for population health programming [30], which typically occurs at larger, regional spatial scales [31]. For example, Chaix et al. identified differing spatial distributions and cluster resolutions of psychoactive substance use and neurotic disorders in Malmö, Sweden, which were associated with adverse social environments [40]. In addition to identifying potential contextual factors for public health action, the analysis also established appropriate levels for intervention by characterising the spatial scales at which variations in mental health outcomes occur [40]. Likewise, Cheung et al. [41] and Ngamini Ngui et al. [42] have reported spatial heterogeneity in suicide across Australia and Québec, Canada, and conclude that understanding this variation is essential to framing national and regional mental health policy. Spatial analysis has also been instrumental in describing geographic variation in psychological susceptibility and its association with resilience factors after Hurricane Sandy in New York City [43].

The objective of this study was to assess the contribution of walkability to geographic variation in mental health outcomes at spatial scales typically used for population-level health programming, planning, and intervention. It builds on our previous work demonstrating the contribution of area-level walkability to geographic variation in population-levels of total walking and moderate and vigorous-intensity physical activity [30]. Our aims were to: (1) evaluate if area-level walkability was associated with area-level psychosocial distress; (2) describe geographic variation in area-level psychosocial distress; (3) assess the contribution of individual-level factors to geographic variation in area-level psychosocial distress; and (4) quantify the contribution of area-level walkability to geographic variation in area-level psychosocial distress not attributable to person-level characteristics using a population-based cohort living in Sydney, Australia. We hypothesised that (1) areal-level psychosocial distress would be spatially structured, and that (2) at least some of this structure would be attributable to area-level walkability.

2. Materials and Methods

2.1. Study Design and Area

We used a cross-sectional, ecological design to investigate geographic variation in psychosocial distress and its relationship to walkability in the Sydney Statistical Division of New South Wales, Australia [44]. Sydney covers a land area of 12,142 km^2 and had a population of 4.1 million persons living in 1.6 million dwellings at the 2006 Australian Census [45]. Analysis was undertaken at the Australian Census of Population and Housing postal area level to coincide with the finest spatial resolution at which the data custodian provided geographical identifiers for 45 and Up Study cohort members. There were were 260 postal areas in Sydney in 2006 [46] with a median land area of 7.6 km^2, 5304 residential dwellings and 13,090 residents [45]. This land area is equivalent to a radial buffer of 1550 m, and corresponds with the upper level of high-resolution buffers used in individual-level studies for which consistent environment-behaviour associations have been reported [47,48].

2.2. Participants

Participants for this study were drawn from The Sax Institute's 45 and Up Study [49]. The 45 and Up Study is a population-based cohort established to investigate health ageing among persons aged 45 years and over in New South Wales, Australia [49]. Recruitment into the study began in January 2006 and was finalised in December 2009 [50] with a total cohort size of 267,153 or 10% of the New South Wales population aged 45 and over [51]. Potential participants were randomly sampled from the Department of Human Services (formerly Medicare Australia) enrolment database, and included an oversample of persons aged 80 years and over. People living in rural areas were also oversampled, and all residents from remote areas were invited to participate [49]; however, neither of these population subgroups are represented in the Sydney Statistical Division. Selected individuals were mailed an invitation letter, and asked to return a signed, written consent form with their baseline survey via reply-paid mail if they consented to participating in the study [49]. We were provided access to the April 2010 data release comprising 266,848 participants [52], which the data custodian had geocoded to 2006 Australian Standard Geographic Classification Statistical Divisions [44] and postal areas [46]. We limited our analysis to participants geocoded to the Sydney statistical division of New South Wales to coincide with the spatial extent of our study factor.

2.3. Data

Individual-level data comprised self-reported responses to the baseline questionnaire of the 45 and Up Study [49], and were used to derive respondent-level outcomes and covariates. Postal area data included the Sydney Walkability Index [23] and 2006 Index of Relative Socioeconomic Disadvantage [53], which were included as study and covariate factors, respectively.

2.4. Outcome Variable

Psychosocial distress served as the outcome factor in our analysis, and was measured using the Kessler Psychological Distress Scale (Kessler 10) [54]. The Kessler 10 is a dimensional measure of non-specific psychosocial distress developed to discriminate between cases and non-cases of serious mental illness in community populations [54]. The scale comprises 10 questions that ask respondents to rate how frequently over the past four weeks they felt tired for no good reason; nervous; so nervous that nothing could calm them down; hopeless; restless or fidgety; so restless that they could not sit still; depressed; that everything was an effort; so sad that nothing could cheer them up; and worthless [54]. Item responses are scored from 1 (none of the time) to 5 (all of the time) and then summed to give a total between 10 and 50. In Australia, scores of 22–29 and ≥30 are sensitive and specific for high and very high levels psychosocial distress in community populations, respectively [55]; specific for any current anxiety or affective disorder [56]; and associated with other mental disorder categories, and presence of any current mental disorder [56]. We created a single, binary outcome variable and classified individuals with a total scale score ≥22 as having high (or very high) psychosocial distress for consistency with existing state and national representative surveys monitoring population levels of psychosocial distress [55,57–59]. List-wise exclusions due to incomplete item responses were minimised by imputing invalid and missing data using the pairing up and mean substitution methods implemented in the Household, Income and Labour Dynamics in Australia Survey [60].

2.5. Study Variable

The primary variable of interest for all analyses was postal area walkability, which we measured using the Sydney Walkability Index. [23]. This index is a three-factor index derived using methods and data comparable to the Neighborhood Quality of Life Study (NQLS) and Physical Activity in Localities and Community Environments (PLACE) walkability indexes [8,22], both of which underpin extensive national and international literatures [23]. The Sydney Walkability Index is calculated within a geographical information system using three built environment variables:

1. Residential dwelling density—the number of residential dwellings per square kilometre of residential land use
2. Intersection density—the number of intersections with three or more roads per square kilometre of total land area
3. Land use mix—the entropy of residential, commercial, industrial, recreational and other land uses.

Environmental variable values are divided into deciles, scored from 1 (lowest) to 10 (highest), summed to give a total score out of 30, and then divided into quartiles corresponding to low, low-medium,

medium-high and high walkability [23]. We have previously demonstrated the predictive validity of the Sydney Walkability Index for utilitarian walking, and its comparability to four-variable indexes (e.g., [8,22]) found in the research literature [23]. We have also recently reported positive associations between the Sydney Walkability Index and population-levels of sufficient walking and total moderate and vigorous-intensity physical activity to enhance health, and its contribution to geographic variation in physical activity behaviours between postal areas in the Sydney statistical division [30].

2.6. Covariates

We included a number of individual- and area-level correlates of psychosocial distress previously identified for the 45 and Up Study cohort in the research literature [61–74]. Individual-level covariates included sex (male, female); five-year age group at baseline interview (45–49 to 80–84 and ≥85 years); language spoken at home (English, other); educational level (less than secondary school, secondary school graduation, trade or certificate or diploma, university degree); relationship status (partner, no partner); employment status (full-time, part-time, other, not working); health insurance type (private with extras, private without extras, Government health care card, none); smoking status (never, past, current), World Health Organisation body mass category (underweight <18.5 kg/m², normal weight 18.5 to <25.0 kg/m², overweight 25.0 to <30.0 kg/m², obese ≥30.0 kg/m²); moderate and vigorous-intensity physical activity in the previous seven days (0, 1–149, 150–299, ≥300 min); number of chronic conditions ever diagnosed and treated in the previous four weeks (0, 1, 2, 3 or more); and limitations on physical functioning (none, minor, moderate, severe). All data were obtained by self-report. Limitations on physical functioning were measured using the Medical Outcomes Study (MOS) 36-Item Short-Form Health Survey (SF-36) physical functioning scale [75,76], with scores of 0 to <60, 60 to <90, 90 to <100, and 100 classified as none, minor, moderate, and severe respectively. Socioeconomic disadvantage was measured at the postal area level using the 2006 Index of Relative Socioeconomic Disadvantage [53]. This index is a general measure of disadvantage derived by principal components analysis of 2006 Australian Census of Population and Housing Census variables indicative of low socioeconomic status (see [53]).

2.7. Statistical Analysis

Our analysis utilised a two-step approach to model relative prevalence within the study cohort. In the first step, the predicted probabilities ($\hat{Y}_{ij}$) of psychosocial distress were estimated for each person from fixed-effect logistic regression models conditioned on individual-level social, economic and health factors as model covariates. We then summed the predicted probabilities for the j postal areas to obtain the total expected numbers of persons with psychosocial distress in each postal area adjusted for its underlying respondent structure (see [30,77–79]).

In the second step, we used used Bayesian Besag, York and Mollié conditional auto regressive models with Poisson likelihoods to estimate prevalence ratios for each of the j postal areas relative to the study area [80]. Besag, York and Mollié spatial models decompose area-level random effects into local, spatially structured (s_j) and global, unstructured (u_j) variance components [81,82] using:

$$log(\theta_j) = \alpha + x_j\beta + s_j + u_j + log(e_j) \tag{1}$$

where θ_j is the prevalence ratio for the jth postal area; α is the mean prevalence ratio for the study area; x_j and β are optional vectors of ecological explanatory variables and parameter estimates, respectively, and e_j is a model offset representing the expected number of cases in the jth area. The unstructured variance component was given a normal prior with mean 0 and precision τ_u^2, while the spatial variance component used an intrinsic conditional auto regressive prior [81] with mean $\bar{s}_j$ and precision tau_j^2 conditioned on the prevalence in the surrounding k postal areas with contiguous boundaries [81]. The hyper-parameters τ_u^2 and τ_s^2 were used to control the variability of u_j and s_j, and were given Gamma hyper-priors of $\gamma(0.5, 0.0005)$ [83]. We derived expected cases e_j using either the overall prevalence ($e_j = p \times n_j$) for unadjusted models or the sum of the predicted probabilities from stage one ($e_j = \sum \hat{Y}_{ij}$) in the case of models adjusted for individual-level factors (see [30,77–79]).

Our analysis fit six analytic and two sensitivity models. Model 1 (M1) was an unadjusted disease mapping model with offsets proportional to the study area prevalence ($p \times n_j$). Model 2 was also a disease model but with individually-adjusted offset terms from stage 1 models. Models 3–6 were ecological regressions: model 3 added postal area walkability to model 2; model 4 added postal area socioeconomic disadvantage to model 2; and model 5 included individually-adjusted offsets, postal area walkability, and postal area socioeconomic disadvantage. Model 6 tested for effect modification of the relationship between psychosocial distress and walkability by socioeconomic disadvantage. We additionally assessed the sensitivity of our association between walkability and psychosocial distress to excluding physical activity level from fixed-effects models used to adjust spatial regression offset terms for individual-level characteristics. These analyses acknowledge the uncertainty regarding the path between walkability and psychosocial distress. If this were mediated by physical activity, as implied by the possible route suggested by Sturm et al. [9], then adjusting for physical activity may suppress the substantive association between walkability and psychosocial distress. We assessed this possibility by refitting models 2 and 3 after excluding physical activity level from the fixed-effect model used to adjust spatial regression offset terms.

Medians and 95% credible intervals for each model parameter were summarised from the posterior distributions of two Monte Carlo Markov Chains initialised using over-dispersed starting values. We ran each chain for 2.5 million iterations and retained every 250th sample to reduce autocorrelation and improve convergence. We discarded the first half of each chain as burn-in, giving 10,000 samples in total for inference. Autocorrelation plots and the Gelman-Rubin diagnostic [84] were used to confirm the convergence of MCMC chains [85]. All models were fit using unweighted survey data, which produce representative and generalisable relative effect estimates for individual-level analyses [86] and unbiased relative effect estimates for postal area analyses [30] in this cohort.

We used the *Deviance Information Criterion* (DIC) to choose between competing conditional auto regressive models with smaller values taken as evidence for improved fits [87]. We also exponentiated and mapped the linear predictor, and spatial and non-spatial random effects for postal areas to identify variation in excess of that attributable to individual- and area-level factors. We additionally calculated spatial fractions ($\rho = \sigma_s^2 / [\sigma_s^2 + \sigma_u^2]$) from the marginal variances of the random effects to estimate the proportion of residual variation in high psychosocial distress due to unobserved and spatially-structured factors (see [88,89]). All data analysis and mapping was undertaken in R 3.3.2. Fixed effects logistic regressions were evaluated at the 5% alpha level and conditional auto regressive Poisson regressions using 95% credible intervals summarised from posterior distributions.

2.8. Ethical and Data Access Statements

The 45 and Up Study is approved and monitored by the University of New South Wales Human Research Ethics Committee (ref no. HREC 05035/HREC 10186). The present research comprised a sub-study of the Social, Environmental, and Economic Factors Study, which is approved and monitored by the University of Sydney Human Research Ethics Committee (ref no. 10-2009/12187). Details on accessing 45 and Up Study data are available on the The Sax Institute website (www.saxinstitute.org.au/our-work/45-up-study).

3. Results

Complete data were available for 91,142 of 115,153 (79.1%) Sydney respondents residing in 254 of 260 (97.7%) study postal areas. The median number of respondents per postal area was 258, with a minimum of 0, maximum of 3302, and inter-quartile range of 145–441 respondents. Table 1 shows individual characteristics for respondents included in our analysis. Similar to the full 45 and Up Study cohort [52], our sample had similar gender and employment characteristics to the study area but was otherwise younger, more highly educated, less likely to speak a language other than English at home, and more likely to be living with a partner than the Sydney population aged 45 years and over [45].

Table 1. Sample characteristics and prevalence estimates for high psychosocial distress.

Variable	Characteristics		Prevalence	
	N	%	n	%
POSTAL AREA LEVEL				
Walkability				
Low	25,217	27.7	1983	7.9
Low-medium	31,023	34.0	2440	7.9
Medium-high	19,232	21.1	1548	8.0
High	15,670	17.2	1154	7.4
Socioeconomic disadvantage				
Q1—Most	17,153	18.8	2096	12.2
Q2	19,272	21.1	1800	9.3
Q3—Middling	14,833	16.3	1109	7.5
Q4	19,789	21.7	1177	5.9
Q5—Least	20,095	22.0	943	4.7
INDIVIDUAL LEVEL				
Sex				
Male	44,220	48.5	3008	6.8
Female	46,922	51.5	4117	8.8
Age				
45–49	13,480	14.8	1328	9.9
50–54	16,619	18.2	1587	9.5
55–59	16,601	18.2	1367	8.2
60–64	13,611	14.9	938	6.9
65–69	10,093	11.1	536	5.3
70–74	6792	7.5	361	5.3
75–79	4898	5.4	319	6.5
80–84	6432	7.1	435	6.8
85+	2616	2.9	254	9.7
Language spoken at home				
English	77,307	84.8	5230	6.8
Other	13,835	15.2	1895	13.7
Education level				
Less than secondary school	7236	7.9	1176	16.3
Secondary school graduation	26,355	28.9	2267	8.6
Trade, certificate or diploma	28,678	31.5	2044	7.1
University degree	28,873	31.7	1638	5.7
Relationship status				
Partner	68,138	74.8	4457	6.5
No partner	23,004	25.2	2668	11.6
Employment status				
Full-time work	32,578	35.7	2052	6.3
Part-time work	13,122	14.4	996	7.6
Other work	1319	1.4	168	12.7
Not working	44,123	48.4	3909	8.9
Health insurance type				
Private with extras	53,835	59.1	3054	5.7
Private without extras	12,822	14.1	746	5.8
Government health care card	11,656	12.8	1974	16.9
None	12,829	14.1	1351	10.5

Table 1. *Cont.*

Variable	Characteristics		Prevalence	
	N	%	n	%
Smoking status				
Never smoked	53,560	58.8	3662	6.8
Past smoker	31,276	34.3	2366	7.6
Current smoker	6306	6.9	1097	17.4
Body mass category				
Underweight	1247	1.4	177	14.2
Normal weight	35,709	39.2	2467	6.9
Overweight	35,555	39.0	2458	6.9
Obese	18,631	20.4	2023	10.9
Total physical activity				
0 min	5296	5.8	912	17.2
1–149 min	15,102	16.6	1635	10.8
150–299 min	15,675	17.2	1185	7.6
≥ 300 min	55,069	60.4	3393	6.2
Diagnosed chronic conditions				
0	31,050	34.1	1397	4.5
1	36,544	40.1	2487	6.8
2	17,915	19.7	2049	11.4
3 or more	5633	6.2	1192	21.2
Treated chronic conditions				
0	41,261	45.3	2683	6.5
1	29,791	32.7	2217	7.4
2	14,285	15.7	1363	9.5
3 or more	5805	6.4	862	14.8
Limited physical functioning				
None	32,198	35.3	1353	4.2
Minor	24,974	27.4	1169	4.7
Moderate	20,074	22.0	1798	9.0
Severe	13,896	15.2	2805	20.2

N Stratum total, **n** Stratum outcome frequency, % Stratum outcome per cent.

3.1. Walkability

We have previously reported in detail on built environment variables and walkability profiles for Sydney postal areas [30]. Environmental variables increased monotonically for low, low-medium, medium-high and high walkability postal areas: residential density (2.3, 13.4, 19.8 and 46 dwellings per hectare), street network connectivity (3.4, 46.1, 79.5 and 162.5 intersections per square kilometre), and land use mix entropy (0.005, 1.033, 0.056, and 0.134), and walkability was distributed along an east-west gradient with highest concentrations of walkable areas surrounding and north of the Sydney central business district, and lowest concentrations in Western Sydney and the peri-urban fringe [23,30].

3.2. Prevalence of Psychosocial Distress

The within cohort prevalence of high psychosocial distress was 7.8% (7.6–8.0%). Prevalence estimates by postal area characteristics are reported at the top of Table 1. Levels of high psychosocial distress were similar in low, low-medium, and medium-high walkability areas, and slightly lower in high walkability areas. In contrast, prevalence of high psychosocial distress decreased monotonically with decreasing relative socioeconomic disadvantage, and was 2.6 times lower in least versus most disadvantaged areas.

3.3. Spatial Analysis

Map A in Figure 1 reports the smoothed distribution of unadjusted prevalence ratio for high psychosocial distress in Sydney statistical division estimated from model 1. There is strong evidence for clustering of high psychosocial distress with a band of relatively higher prevalence postal areas stretching from the north, through the centre, and then to the south-eastern border of Sydney. Prevalence ratios were consistently lower for postal areas in the central business and surrounding districts on the eastern seaboard, and in south western Sydney. Maps B and C decompose the total prevalence into its spatial and unstructured sources, respectively. Map C indicates that little variation is due to unstructured factors, while map B shows that the distribution of high psychosocial distress is largely attributable to unobserved and spatially-structured factors. This is confirmed by the Model 1 spatial fraction reported in Table 2, which attributes almost all of the variation in map A to the spatial random effect.

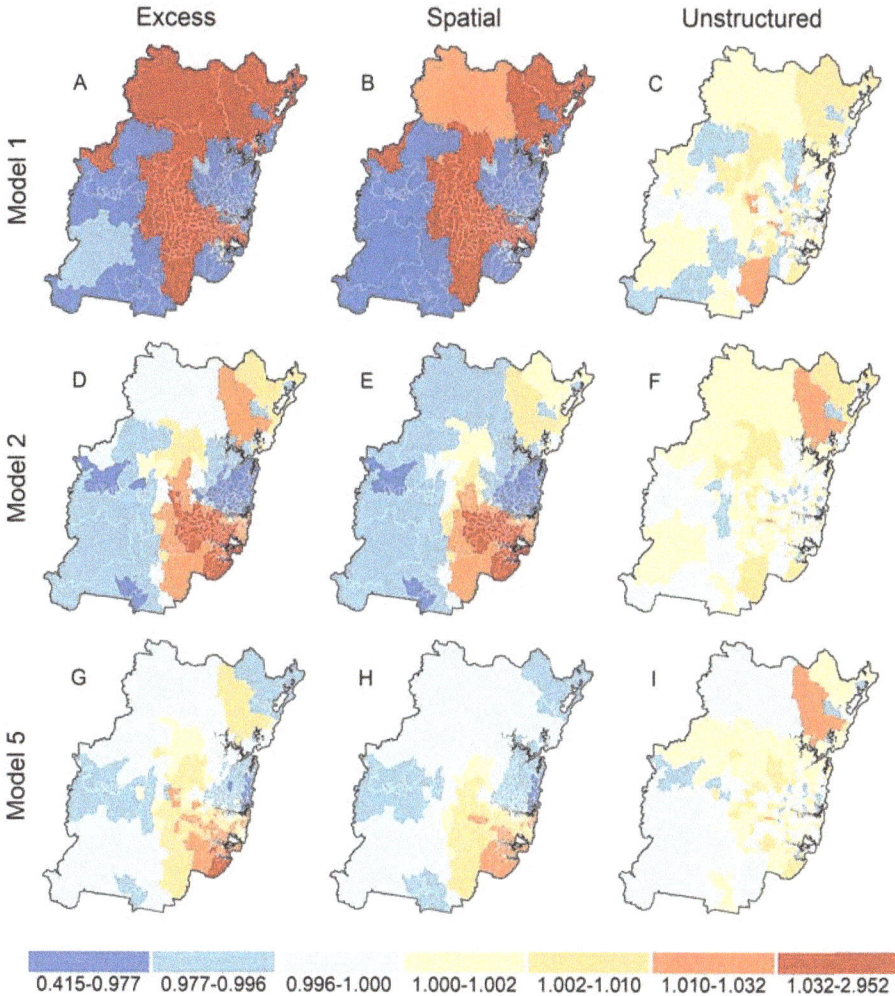

Figure 1. Total, Spatial and Unstructured prevalence ratios for Sydney postal areas. Total prevalence ratios were derived by exponentiating the sum of the log odds for the s and u random effects; Spatial and Unstructured prevalence ratios were obtained by exponentiating the log odds of the individual s and u components, respectively. Total, Spatial, and Unstructured prevalence ratio estimates are reported in maps **A–C** for model 1, maps **D–F** for model 2, and maps **G–I** for model 5.

Table 2. Conditional auto regression model summaries for high psychosocial distress.

Individual-Level Adjustment	Model 1	Model 2	Model 3	Model 4	Model 5
	No	Yes	Yes	Yes	Yes
Prevalence ratios (95% CrI)					
Constant	0.99 (0.96–1.02)	0.99 (0.97–1.02)	0.97 (0.91–1.03)	1.07 (1.02–1.12)	1.04 (0.97–1.12)
Walkability					
Low			1.00		1.00
Low-medium			1.01 (0.94–1.08)		1.00 (0.94–1.07)
Medium-high			1.08 (0.99–1.18)		1.07 (0.99–1.16)
High			1.03 (0.93–1.15)		1.03 (0.94–1.13)
Socioeconomic disadvantage					
Q1—Most				1.00	1.00
Q2				0.98 (0.91–1.04)	0.98 (0.91–1.05)
Q3— Middling				0.92 (0.86–1.00)	0.92 (0.85–0.99)
Q4				0.90 (0.83–0.98)	0.90 (0.83–0.97)
Q5—Least				0.82 (0.76–0.90)	0.83 (0.76–0.90)
Model diagnostics					
pD	127.85	21.73	24.40	15.32	17.20
DIC	1557.25	1418.33	1419.26	1409.06	1410.40
Spatial fraction	0.99	0.88	0.88	0.61	0.55

CrI credible interval, **pD** effective parameters, **DIC** Deviance Information Criterion. **Model 1** null model with expected cases proportional to the overall prevalence. **Model 2** null model with expected cases adjusted for individual-level factors. **Model 3** Model 2 + Sydney Walkability Index. **Model 4** Model 2 + Index of Relative Socioeconomic Disadvantage. **Model 5** Model 3 + Index of Relative Socioeconomic Disadvantage.

Table 3 reports unadjusted odds ratios (OR) for associations between high psychosocial distress and individual-level covariates, which were used to adjust expected values in spatial models. All variables were statistically significant and important in univariate models with small to medium effect sizes [90]. Odds for high psychosocial distress were increased for females, people who spoke a language other than English at home, had less than a university education, were not working full-time, did not have private health insurance, or were on a government health care card. Higher odds were also observed for current and past smokers, persons who were underweight or obese, had one or more chronic conditions ever diagnosed or treated in the last month, or experienced minor to severe physical limitation. Reduced odds of high psychosocial distress were associated with older age, peaking in ages 65–74, and longer durations of total moderate and vigorous-intensity physical activity per week.

Table 3. Unadjusted and fully-adjusted odds ratios for individual-level adjustment variables.

	Unadjusted		Adjusted	
	OR	95% CI	OR	95% CI
Sex	*p* < 0.0001		*p* = 0.2434	
Male	1.00		1.00	
Female	1.32	1.25–1.38	0.97	0.91–1.02
Age	*p* < 0.0001		*p* < 0.0001	
45–49	1.00		1.00	
50–54	0.97	0.89–1.04	0.82	0.76–0.89
55–59	0.82	0.76–0.89	0.57	0.52–0.62
60–64	0.68	0.62–0.74	0.36	0.32–0.39
65–69	0.51	0.46–0.57	0.21	0.18–0.24
70–74	0.51	0.46–0.58	0.16	0.14–0.18
75–79	0.64	0.56–0.72	0.16	0.14–0.19
80–84	0.66	0.59–0.74	0.13	0.12–0.15
85+	0.98	0.85–1.13	0.14	0.12–0.17
Language spoken at home	*p* < 0.0001		*p* < 0.0001	
English	1.00		1.00	
Other	2.19	2.07–2.31	1.92	1.80–2.04
Education level	*p* < 0.0001		*p* < 0.0001	
Less than secondary school	3.23	2.98–3.50	1.70	1.55–1.87
Secondary school graduation	1.56	1.47–1.67	1.20	1.12–1.29
Trade, certificate or diploma	1.28	1.19–1.36	1.09	1.02–1.18
University degree	1.00		1.00	
Relationship status	*p* < 0.0001		*p* < 0.0001	
Partner	1.00		1.00	
No partner	1.87	1.78–1.97	1.41	1.33–1.50
Employment status	*p* < 0.0001		*p* < 0.0001	
Full-time work	1.00		1.00	
Part-time work	1.22	1.13–1.32	1.14	1.05–1.24
Other work	2.17	1.84–2.57	1.57	1.30–1.89
Not working	1.45	1.37–1.53	1.46	1.35–1.58
Health insurance type	*p* < 0.0001		*p* < 0.0001	
Private with extras	1.00		1.00	
Private without extras	1.03	0.95–1.12	1.03	0.94–1.12
Government health care card	3.39	3.19–3.60	1.78	1.65–1.92
None	1.96	1.83–2.09	1.36	1.27–1.47

Table 3. *Cont.*

	Unadjusted		Adjusted	
	OR	95% CI	OR	95% CI
Smoking status	$p < 0.0001$		$p < 0.0001$	
Never smoked	1.00		1.00	
Past smoker	1.12	1.06–1.18	1.07	1.00–1.13
Current smoker	2.87	2.67–3.09	1.64	1.51–1.78
Body mass category	$p < 0.0001$		$p < 0.0001$	
Underweight	2.23	1.89–2.63	1.61	1.34–1.93
Normal weight	1.00		1.00	
Overweight	1.00	0.94–1.06	0.93	0.87–0.99
Obese	1.64	1.54–1.75	0.88	0.82–0.94
Total physical activity	$p < 0.0001$		$p < 0.0001$	
0 min	1.00		1.00	
1–149 min	0.58	0.53–0.64	0.75	0.68–0.82
150–299 min	0.39	0.36–0.43	0.64	0.58–0.71
≥ 300 min	0.32	0.29–0.34	0.58	0.53–0.64
Diagnosed chronic conditions	$p < 0.0001$		$p < 0.0001$	
0	1.00		1.00	
1	1.55	1.45–1.66	1.56	1.45–1.68
2	2.74	2.55–2.94	2.45	2.26–2.66
3 or more	5.70	5.24–6.19	4.32	3.90–4.78
Treated chronic conditions	$p < 0.0001$		$p < 0.0240$	
0	1.00		1.00	
1	1.16	1.09–1.23	1.02	0.96–1.10
2	1.52	1.42–1.62	1.01	0.93–1.10
3 or more	2.51	2.31–2.72	1.17	1.05–1.29
Limited physical functioning	$p < 0.0001$		$p < 0.0001$	
None	1.00		1.00	
Minor	1.12	1.03–1.21	1.24	1.15–1.35
Moderate	2.24	2.09–2.41	2.15	1.98–2.33
Severe	5.77	5.38–6.17	4.41	4.05–4.79

OR Odds ratio, **CI** Confidence interval

Adjusted OR remained important but were attenuated relative to unadjusted effect estimates (see Table 3). The two exceptions were age and body mass category. The protective effect of age relative to persons 45–49 year became stronger throughout the life span following adjustment, peaking in the 80–84 years age group, while odds of high psychosocial distress for obese relative to normal weight individuals switched from 1.64 (1.54–1.75) to 0.88 (0.82–0.94). The latter was due to confounding of the association by limitations on physical functioning, age, and number of chronic conditions ever diagnosed. Obese respondents with high psychosocial distress were more likely to have severe functional limitations (50.2% versus 35.1%) or been diagnosed with three or more chronic health conditions (24.8% versus 13.5%), and less likely to be aged 80 years or older (4.9% versus 11.5%) compared to non-obese persons.

The second row of maps in Figure 1 shows relative prevalence of high psychosocial distress (map D), decomposed into to spatially structured (map E) and unstructured (map F) factors after accounting for individual-level differences between Sydney postal areas (model 2). The magnitude of prevalence ratios were substantially attenuated and reduced in range from 0.42–2.92 for model 1 to 0.86–1.09 for model 2. Despite this reduction, prevalence ratio remained geographically clustered with higher rates in central and south-eastern Sydney, and lower rates in north Sydney (see maps

D–F). The DIC and pD for model 2 indicated a substantially better fit over model 1, which reduced spatial and unstructured variation by 98.5% and 52.1%, respectively, and the spatial fraction by 11.1% (see Model 2 in Table 2).

Associations between high psychosocial distress and postal area walkability (model 3) and relative socioeconomic disadvantage (model 4) are reported in Table 2. We found no evidence for an association between psychosocial distress and postal area walkability after adjusting for individual-level factors. The DIC and pD for model 3 indicated a poorer fit compared to model 2, and all walkability credible intervals included unity. Excluding physical activity level from model offsets in sensitivity analyses did not alter prevalence ratios obtained from model 3 (see Table 4). The increase in DIC (0.98) and pD (2.39) for this sensitivity model relative to a baseline sensitivity model excluding walkability and physical activity also provided no support for an association between psychosocial distress and walkability, or excluding physical activity from our analysis (see Table 4). Model 4 added relative socioeconomic disadvantage to model 2, which also included individual-level socioeconomic factors, resulted in an improved model that reduced DIC by 9.3 units and pD by 6.4 parameters. Compared to postal areas in the most socioeconomically disadvantaged quintile 1, prevalence of high psychosocial distress was similar for postal areas in quintile 2, and 8%, 10% and 18% lower for postal areas in quintiles 3–5, respectively (see Table 2).

The bottom row of Figure 1 displays final prevalence ratios from model 5 for high psychosocial distress (map G) decomposed into spatially structured (map H) and unstructured (map I) factors after accounting for individual differences, and postal area walkability and socioeconomic disadvantage. Simultaneously adjusting for individual and postal area factors further attenuated prevalence ratios but did not substantially affect the geographic distribution of high psychosocial distress, which remained higher-than-expected in central and south-eastern Sydney, and lower-than-expected in north Sydney. Adjusting for relative socioeconomic disadvantage in model 5 did not alter effect estimates or conclusions for the association between postal area walkability and high psychosocial distress from model 3 (see Table 2). The DIC value for model 5 was 1.3 units larger than the "best" fitting model 4 but within the ≤2 unit change range indicating a model deserving consideration [87]. Spatial and unstructured variation in fully adjusted model 5 were reduced by 99.9% and 59.1% relative to unadjusted model 1, and the spatial fraction reduced from 0.99 to 0.55 (38.4%). Interaction model 6 provided no evidence that the association between walkability and high psychosocial distress was modified by postal area socioeconomic disadvantage ($DIC_{M6} - DIC_{M5} = 18.1$).

Table 4. Conditional auto regression model summaries for sensitivity analyses.

	Baseline	Walkability
Prevalence ratios (95% CrI)		
Constant	0.97 (0.97–1.02)	0.97 (0.91–1.03)
Walkability		
Low	–	1.00
Low-medium	–	1.01 (0.94–1.08)
Medium-high	–	1.08 (0.99–1.18)
High	–	1.03 (0.93–1.15)
Model diagnostics		
pD	23.58	25.97
DIC	1420.05	1420.99
Spatial fraction	0.90	0.90

CrI credible interval, **pD** effective parameters, **DIC** Deviance Information Criterion. **Baseline** null model with adjusted offsets EXCLUDING individual physical activity level. **Walkability** Baseline + Sydney Walkability Index.

4. Discussion

This appears to be the first study to assess associations between area-level walkability and psychosocial distress using a large population cohort within a spatial framework. Our findings indicate that while psychosocial distress is geographically clustered in the Sydney statistical division, area-level walkability does not contribute to this spatial structure, which is principally patterned by the individual-level characteristics of residents within postal areas. We did, however, observe a consistent association between postal area socioeconomic disadvantage and prevalence of high psychosocial distress independent of individual-level social and economic factors. Prevalence of high psychosocial distress is 10–18% lower in the least compared to most socioeconomically disadvantaged postal areas after adjusting for individual-level differences and postal area walkability. Our results suggest that while area-level socioeconomic disadvantage makes a small contribution to geographic variation in psychosocial distress (2.2%), programming and planning activities will likely deliver greatest benefits by focusing on individual-level determinants, correlates, and mediators of disease burden and inequality associated with psychosocial distress.

Modifying the walkability of built environments to improve the health of populations is frequently recommended [8,23,91–95], and has been suggested as a potential focus for community-level mental health planning [13]. Such recommendations implicitly assume that individual-level environment-behaviour and environment-outcome findings scale to community- and population levels. However, these assumptions are rarely evaluated, which leaves open the potential for spurious cross-level action due to atomistic [96] or individualistic [97] fallacy. Our study is novel in that we have directly examined associations between area-level walkability and high psychosocial distress in Sydney at spatial scales more typical of population-level programming, planning, and intervention. At these scales, we observed substantial geographic variation in unadjusted disease maps of psychosocial distress prevalence for postal areas. However, we found no evidence supporting a link between walkability and prevalence of psychosocial distress or its geographic patterning, both of which appear largely attributable to the spatial distribution of individual-level factors across the Sydney statistical division with a small contribution from postal area socioeconomic disadvantage.

An evidence base linking walkability to mental health outcomes is only beginning to emerge in the research literature, and is presently derived from a small number of individual-level studies. Berke et al. reported in 2007 that the odds of depression in the Adult Changes in Thought (ACT) Study cohort were reduced by a factor of 0.31–0.33 for the highest versus lowest walkability quartile but only for older men. In contrast, a 2011 cross-sectional study of older men in Perth, Australia, found that while depression was unrelated to Census Collection District walkability, it was associated with individual environmental variables used to construct their index, with increased odds of depression in Districts with middling (1.10–2.16) and high (1.08–2.14) versus low land use mix, and some versus no retail land use (1.04–1.90) [10]. However, a subsequent study of older Welsh men has reported reduced odds of psychosocial distress for greater land use mix (0.22–1.00) and street network connectivity (0.28–1.00) [11], another built environment variable routinely included in walkability indexes [8,22]. While most recently, James et al have reported that the odds of depression (1.08–1.16) and current anti-depressant use (1.08–1.25) were significantly increased among persons living in the highest versus least walkable neighbourhoods of low-income and racially diverse populations in south eastern United States [12].

The heterogeneity of findings from these studies likely reflects the considerable variability in methods and measures they employed [19]. Walkability was assessed using both objective and perceived methods, and no two studies used the same index, scale or combination of environmental variables to measure walkability. Likewise, mental health outcomes were assessed for a diverse range of conditions and symptoms using a mix of standardised scales and self-report. In their 2016 systematic review, Gong et al. identified an urgent need to develop standardised approaches to researching built environment influences on mental health [19]. This concern reflects a broader focus in the walkability literature to reconcile environment-behaviour research methods to improve between-study comparability and inform public health policy and planning (e.g., [21,23,98]). Gong et al. have also emphasised the importance of utilising objective built environment indexes in mental health research to reduce information bias resulting from a tendency among persons with poorer mental health to perceive their environments more negatively [19]. Our study design is consistent with these recommendations in its use of validated outcome and objective exposure variables, which are routinely used for population-level health surveillance [34], and individual- and area-level walkability research (e.g., [8,22,23,95,99]).

The mechanism by which walkability my influence psychosocial distress remains an important but unresolved issue for current and future environmental exposure research. Both physical activity [9] and social capital [17] have been hypothesised as plausible variables through which environmental walkability acts on mental health; however, neither has been evaluated within a causal framework. Our study indirectly considered the influence of individual-level physical activity on area-level associations between walkability and psychosocial distress through a sensitivity analysis that excluded physical activity from the model used to derive offset terms for ecological regressions. We obtained the same effect estimates for environmental walkability regardless of whether we adjusted for physical activity or not; however, our study design (cross-sectional) and analytic approach (ecological) preclude us from making inferences about the possible mediating role of this variable. Ideally, any evaluation of potentially mediating variables should use prospectively collected data from multiple waves of follow-up to allow sufficient time to elapse between the hypothesised cause and its effect, and to avoid the bias that arises when cross-sectional data are used to estimate longitudinal effects [100]. The 45 and Up Study comprises 265,000 persons aged 45 years and older [49], with 40% residing in a geographical unit classified by the Sydney Walkability Index. Follow-up of this cohort occurs approximately quinquennially, with a third wave of data collection scheduled to begin in the next few years. This will provide a unique opportunity to evaluate potential causal pathways between walkability and mental health, and how they may contribute to healthy ageing.

We observed strong associations between all individual-level socioeconomic indicators and psychosocial distress. This is consistent with the substantive (e.g., [101–105]) and 45 and Up Study literatures (e.g., [68]) indicating higher prevalence of poorer mental health in more socioeconomically disadvantaged individuals, regardless of how mental health and socioeconomic status are measured. Odds of psychosocial distress were 1.05–1.89, 1.27–1.92, and 1.02–1.87 times higher for persons not in full-time work, without private health insurance, and without a university degree, respectively. We also observed a consistent contextual effect of relative socioeconomic disadvantage on prevalence of psychosocial distress that reduced postal area ratios by 10–24% in the least compared to most disadvantaged quintiles. This gradient is supported by a recent narrative review, which reported consistent evidence for a contextual socioeconomic effect over-and-above that due to individual-level socioeconomic factors [106]. In our study, this contextual effect accounts for approximately 2.3% of the spatial and 4.2% of the non-spatial variation in prevalence of postal area psychosocial distress that remains after adjusting for individual-level factors and area-level walkability. This is smaller than the 13.5% of residual unstructured variation in depression prevalence from World Health Organisation health surveys due to country-level income and income inequality reported by Rai et al. [107]; similar to the 4.5% of unstructured variation in depressive symptoms due to area-level mean income and Gini Coefficient reported by Lee at al. for 253 Korean communities [108]; and consistent with review evidence indicating individual-level factors account for most of the unstructured variation between higher-order cluster units [106]. Cross-level interactions between area- and individual-level socioeconomic status were beyond the scope of this study; however, the available evidence suggests that poorer individual-level socioeconomic position increases susceptibility to neighbourhood-level socioeconomic disadvantage, while improved individual-level position buffers against this effect [106].

We also observed very strong associations between psychosocial distress and numbers of chronic conditions ever diagnosed, and psychosocial distress and limitations on physical functioning. The odds of psychosocial distress were 1.6, 2.5, and 4.3 times higher for person with 1, 2 or 3 or more doctor diagnosed chronic conditions than those with none. Similarly, the odds of psychosocial distress among respondents with minor, moderate or severe limitations on physical functioning were 1.2, 2.2, and 4.4 times higher than those with none. These findings agree with previously published studies on the correlates of psychosocial distress and depression among 45 and Up Study participants [64,65], and Australian [109] and international [110] primary care cohorts. Ormel et al. have identified three components to associations between depressive symptoms and functional disability: an immediate effect of decreased physical function on depressive symptoms; a weaker, lagged effect of functional disability leading depressive symptoms; and a weak, lagged effect of depressive symptoms leading functional disability, all of which may be modified by personal resilience factors and access to effective care [111]. The symptoms of depression and psychosocial distress might also be exacerbated by the social stresses and stigma associated with reduced physical function [65]. These possibilities are consistent with reports that psychosocial distress is more strongly related to level of disability among 45 and Up Study participants with cancer than the fact of a cancer diagnosis [62]. Our study employed a cross-sectional design, which precluded us from identifying the directionality of associations between psychosocial distress, multiple chronic diseases, and limitations on physical functioning. However, our findings do support a role for these factors in the geographical patterning of psychosocial distress across the Sydney statistical district, which is likely to be especially informative for planners, policy-makers, and researchers for population-level health programming, intervention, and evaluation activities.

Although we found no evidence for an association between postal area walkability and psychosocial distress, our findings still have relevance for population-level mental health planning. First, our study demonstrates the utility of visualising geographic variation in mental health outcomes to identify areas with higher or lower than expected rates, which may provide targets for population-level intervention. The utility of mapping for service planning has been demonstrated by Bazemore et al., who used geographical information systems to visualise and address discrepancies between services delivered and under-served areas in a North American primary care clinic network [112]. Our findings indicate that psychological distress is geographically clustered in Sydney, and that it is the spatial distribution of individual-level demographic, social, economic and health factors that drive this patterning. From a planning perspective, adding or removing individual-level factors sequentially and visualising their effect on disease maps would be especially informative for identifying those individual-level characteristics and circumstances contributing to higher-than-expected psychosocial distress in a specific geographic area. This was beyond the scope of our research, which was concerned with the contribution of area-level walkability to postal area psychosocial distress prevalence and geographic variation over and above that attributable to individual level factors. We observed no association between postal area psychosocial distress and walkability, and walkability had little effect on disease maps. This suggests area-level walkability is insufficiently sensitive for informing population health policy and programming aimed at improving mental through built environment intervention, and that planners and policy-makers are more likely to maximise health gains by focusing on established individual-level correlates and determinants of mental (ill) health.

A major strength of our study is it use of the large, high-quality 45 and Up Study cohort, which has population-level coverage. However, similar to the larger cohort, our sample was younger, better educated, and more likely to be partnered and speak English at home than the general population aged 45 years and over in the Sydney statistical district. While this precludes us from generalising point-prevalence estimates beyond our sample, it is likely that our relative effect estimates are externally valid. It is well established in the epidemiological literature that relative measures of risk and odds derived from cohorts are usually generalisable irrespective of representativeness and non response [113,114]. This has been specifically demonstrated in the case of the 45 and Up Study by Mealing et al., whom reported that odds ratio estimates from this cohort are highly comparable to those derived from the population-representative New South Wales Continuous Health Survey [86]. We have also reported very high correlations between postal area relative risks and disease maps estimated from unweighted and post-stratification weighted data, which indicates spatial risk estimators within the 45 and Up Study cohort are unaffected by non response bias [30,115].

Our study used validated measures for quantifying outcomes and exposures. The Kessler 10 [54] is an established, scale-derived measure of psychosocial (psychological) distress that is routinely used in research and to monitor mental health status in population-representative surveys [34], making it an ideal choice for our application. Similarly, the Sydney Walkability Index is an established indicator of the built environment with demonstrated validity and specificity for walking behaviour at a range of spatial scales [23,30]. Objectively characterising the walkability of built environments is especially important for mental health outcomes where systematic information bias is reasonably expected [19]. In addition to being objective, the Sydney Walkability Index is constructed using the same methods as other influential indexes in the walkability literature (see [8,22]). However, while our index is derived at the area-level, we caution against interpreting it as a proxy for individual-level exposure [30]. We deliberately matched the spatial scales at which we measured outcome and exposure variables to avoid validity concerns arising from cross-level inference [116], which was evident in at least one of the individual-level studies reviewed (see [10]). We argue that when walkability exposure and outcome are measured at the same area-level resolution, it constitutes a contextual variable describing the shared walkability experience of populations and groups inhabiting the same geographic space [30]; we have demonstrated the plausibility of this conceptualisation using the same cohort and spatial scale (see [30]). We believe this makes our approach especially relevant to planning applications, which typically occur at regional levels and for populations of individuals.

Another strength of our study is its use Bayesian Besag, York and Mollié spatial models fit as disease mapping and ecological regressions to: (1) directly assess associations between outcomes, exposures, and covariates; (2) quantify geographic clustering of high psychosocial distress; and (3) evaluate the contribution of postal area walkability to this spatial structure. Spatial methods are increasingly employed in the epidemiological literature to understand the role of place on health outcomes, behaviours and determinants, and to account for spatial autocorrelation, which is problematic for valid inference if not handled appropriately [117]. Our study demonstrates the highly spatial nature of psychosocial distress in Sydney and the importance of handling this geographic structure at the analysis stage. While standard multilevel analysis can account for autocorrelation through random effect terms, our study highlights the advantage of decomposing this variation into spatial and non-spatial sources for informing programming, planning, and intervention activities. We also avoided potential confounding in our analysis due to individual differences in the underlying response populations by adjusting model offsets using predicted probabilities from individual-level fixed-effects regressions of psychosocial distress on person-level demographic, social, economic and health factors. This approach is commonly employed in the epidemiological literature to adjust area-level models where individual-level variables cannot be parameterized within a parsimonious model [78] or would be computationally prohibitive [30,77,79].

Our study is subject to a number of limitations. We were unable to include a measure of social social capital in our study despite its hypothesised link with walkability and psychosocial distress. Self-reported measures of social capital were collected as part of the 45 and Up Study baseline survey but were poorly completed. Limiting our analysis to cases with complete data on these variables would have further reduced our effective sample size, and resulted in a non-response rate well above the maximum 20% identified for cohort studies and data that are missing not at random (MNAR) (see [118,119]). However, we do not believe including social capital would have substantially altered our findings for two reasons. First, we observed no association between walkability and psychosocial distress for social capital to be considered a potential mediator [120]. And second, a recent individual-level study of the association between walkability and mental health reported that effect estimates were unchanged when social capital was included in statistical models [12], which is inconsistent with a moderating effect by social capital [121].

Another limitation of our study is that individual- and area-level factors were modelled separately. Ideally, all variables would be included in a single, parsimonious model that allowed their joint effects to be assessed concurrently. These types of multi-level spatial models are beginning to emerge in the epidemiological literature (e.g., [122]) but are not easily implemented in standard statistical software, and are often computationally prohibitive for problems with large sample sizes and numerous spatial units outside of high performance computing environments [122]. Our approach to adjusting spatial models using offset terms derived from fixed-effect analyses of individual-level factors is commonly employed in the epidemiological literature where a parsimonious model cannot be specified or is computationally prohibitive [30,77–79], as was the case in this study. However, recent methodological advances incorporating Integrated Nested Laplace Approximation (INLA) to estimate approximate posteriori marginals appear to offer a potential solution for the efficient fitting of these multi-level Bayesian spatial models [123,124].

Finally, our study used Australian-specific postal areas as the units of analysis, and sample-specific cut-points for the calculation of Sydney Walkability Index variables. The spatial extents of postal areas may not coincide with the planning units used in other jurisdictions. Associations between outcomes and exposures can vary with geographic resolution, even when both are measured at the same spatial scale [125]. As such, this should be taken into consideration when applying our findings at finer or coarser spatial scales. However, we do note that the median land area of our postal areas was 7.6 km^2 or the equivalent of a 1550 m radial buffer, which is at the upper limit of buffer sizes used in individual-level studies, and for which consistent environment-behaviour associations have been reported [47,48]. We also quantized environmental variables relative to their distribution in the Sydney statistical division, which may not be representative of other jurisdictional spatial units. To address this potential limitation we have reported the cut-points used to construct our index [30], and encourage planners, policy-makers, and researchers to use these in assessing the applicability of our results to their setting of interest. We also acknowledge that the cross-sectional design of our study limits its conclusions to non-causal inferences.

5. Conclusions

Walkability describes the capacity of the built environment to promote or hinder walking for multiple purposes, and has been proposed by Berke and colleagues as a potential environmental focus for mental health planning and intervention [13]. Our study examined this possibility at a spatial scale similar to those typically used for regional-level planning and found no evidence for an association between postal area walkability and high psychosocial distress in the Sydney Statistical Division that could be leveraged for this purpose. We did, however, observe strong geographic clustering of high psychosocial distress, which was largely attributable to individual-level factors with a small contribution from area-level socioeconomic disadvantage. These findings suggests that mental health planning and intervention activities will likely deliver greatest benefits by focusing on individual-level determinants, correlates, and mediators of disease burden and inequality associated with psychosocial distress and other mental health outcomes.

Acknowledgments: This research was completed using data collected through the 45 and Up Study (www.saxinstitute.org.au). The 45 and Up Study is managed by the Sax Institute in collaboration with major partner Cancer Council NSW; and partners: the National Heart Foundation of Australia (NSW Division); NSW Ministry of Health; NSW Government Family & Community Services—Ageing, Carers and the Disability Council NSW; and the Australian Red Cross Blood Service. We thank the many thousands of people participating in the 45 and Up Study. Details on accessing 45 and Up Study data are available on the The Sax Institute website (www.saxinstitute.org.au/our-work/45-up-study). We wish to acknowledge Associate Professor Philayrath Phongsavan for her tireless and effective coordination of this project. DJM is grateful to Hevan Corrimal, WHC Sutherland, and OCB Wollongong for their ongoing support and providing office space to undertake this research. This study is part of the "Understanding the impact of the social, economic and environmental factors on the health of Australians in mid—later life; where are the opportunities for prevention?" study (National Health & Medical Research Council (NHMRC) Grant 402810). We are grateful to the two anonymous reviewers who provided helpful comments on our manuscript.

Author Contributions: D.J.M. undertook all statistical analyses and wrote the first draft of the manuscript with contributions from G.G.M., A.E.B., and B.B.J. All authors were involved in study conception, and contributed to and approved the final manuscript.

Conflicts of Interest: The authors declare no conflict of interest. This manuscript was reviewed for technical accuracy by The 45 and Up Study coordinating centre prior to its submission for peer review. Neither the 45 and Up Study coordinating centre nor the NHMRC had a role in the design of the study; the collection, analysis, or interpretation of data; the writing of the manuscript; or in the decision to publish the results.

References

1. Vigo, D.; Thornicroft, G.; Atun, R. Estimating the true global burden of mental illness. *Lancet Psychiatry* **2016**, *3*, 171–178.

2. Vos, T.; Abajobir, A.A.; Abate, K.H.; Abbafati, C.; Abbas, K.M.; Abd-Allah, F.; Abdulkader, R.S.; Abdulle, A.M.; Abebo, T.A.; Abera, S.F.; et al. Global, regional, and national incidence, prevalence, and years lived with disability for 328 diseases and injuries for 195 countries, 1990–2016: A systematic analysis for the Global Burden of Disease Study 2016. *Lancet* **2017**, *390*, 1211–1259.

3. Hay, S.I.; Abajobir, A.A.; Abate, K.H.; Abbafati, C.; Abbas, K.M.; Abd-Allah, F.; Abdulkader, R.S.; Abdulle, A.M.; Abebo, T.A.; Abera, S.F.; et al. Global, regional, and national disability-adjusted life-years (DALYs) for 333 diseases and injuries and healthy life expectancy (HALE) for 195 countries and territories, 1990–2016: A systematic analysis for the Global Burden of Disease Study 2016. *Lancet* **2017**, *390*, 1260–1344.

4. World Health Organization. *Depression and Other Common Mental Disorders: Global Health Estimates*; World Health Organization: Genva, Switzerland, 2017.

5. Slade, T.; Johnston, A.; Teesson, M.; Whiteford, H.; Burgess, P.; Pirkis, J.; Saw, S. *The Mental Health of Australians 2: Report on the 2007 National Survey of Mental Health and Wellbeing*; Department of Health and Ageing: Canberra, Australia, 2009.

6. World Health Assembly. Global Burden of Mental Disorders and the Need for a Comprehensive, Coordinated Response From Health and Social Sectors at the Country Level: Report by the Secretariat (A65/10 Item 13.2). Available online: http://www.who.int/iris/handle/10665/78898 (accessed on 31 January 2018).

7. World Health Organization. *Mental Health Action Plan 2013–2020*; World Health Organization: Geneva, Switxerland, 2013.

8. Leslie, E.; Coffee, N.; Frank, L.; Owen, N.; Bauman, A.; Hugo, G. Walkability of local communities: Using geographic information systems to objectively assess relevant environmental attributes. *Health Place* **2007**, *13*, 111–122.

9. Sturm, R.; Cohen, D.A. Suburban sprawl and physical and mental health. *Public Health* **2004**, *118*, 488–496.

10. Saarloos, D.; Alfonso, H.; Giles-Corti, B.; Middleton, N.; Almeida, O.P. The built environment and depression in later life: The health in men study. *Am. J. Geriatr. Psychiatry* **2011**, *19*, 461–470.

11. Sarkar, C.; Gallacher, J.; Webster, C. Urban built environment configuration and psychological distress in older men: Results from the Caerphilly study. *BMC Public Health* **2013**, *13*, 695.

12. James, P.; Hart, J.E.; Banay, R.F.; Laden, F.; Signorello, L.B. Built Environment and Depression in Low-Income African Americans and Whites. *Am. J. Prev. Med.* **2017**, *52*, 74–84.

13. Berke, E.M.; Gottlieb, L.M.; Moudon, A.V.; Larson, E.B. Protective association between neighborhood walkability and depression in older men. *J. Am. Geriatr. Soc.* **2007**, *55*, 526–533.

14. Physical Activity Guidelines Advisory Committee. Physical Activity Guidelines Advisory Committee Report. 2008. Available online: http://health.gov/paguidelines/guidelines/ (accessed on 31 January 2018).

15. Josefsson, T.; Lindwall, M.; Archer, T. Physical exercise intervention in depressive disorders: Meta-analysis and systematic review. *Scand. J. Med. Sci. Sports* **2014**, *24*, 259–272.

16. Rebar, A.L.; Stanton, R.; Geard, D.; Short, C.; Duncan, M.J.; Vandelanotte, C. A meta-meta-analysis of the effect of physical activity on depression and anxiety in non-clinical adult populations. *Health Psychol. Rev.* **2015**, *9*, 366–378.

17. Leyden, K.M. Social capital and the built environment: The importance of walkable neighborhoods. *Am. J. Public Health* **2003**, *93*, 1546–1551.

18. Porta, M.; Greenland, S.; Hernán, M.; dos Santos Silva, I.; Last, J.M. *A Dictionary of Epidemiology*, 6th ed.; Oxford University Press: Oxford, UK, 2014.

19. Gong, Y.; Palmer, S.; Gallacher, J.; Marsden, T.; Fone, D. A systematic review of the relationship between objective measurements of the urban environment and psychological distress. *Environ. Int.* **2016**, *96*, 48–57.
20. Ehsan, A.M.; De Silva, M.J. Social capital and common mental disorder: A systematic review. *J. Epidemiol. Community Health* **2015**, *69*, 1021.
21. Saelens, B.E.; Handy, S.L. Built environment correlates of walking: A review. *Med. Sci. Sports Exerc.* **2008**, *40*, S550–S566.
22. Frank, L.D.; Sallis, J.F.; Saelens, B.E.; Leary, L.; Cain, K.; Conway, T.L.; Hess, P.M. The development of a walkability index: Application to the Neighborhood Quality of Life Study. *Br. J. Sports Med.* **2010**, *44*, 924–933.
23. Mayne, D.; Morgan, G.; Willmore, A.; Rose, N.; Jalaludin, B.; Bambrick, H.; Bauman, A. An objective index of walkability for research and planning in the Sydney metropolitan region of New South Wales, Australia: An ecological study. *Int. J. Health Geogr.* **2013**, *12*, 61.
24. Durand, C.P.; Andalib, M.; Dunton, G.F.; Wolch, J.; Pentz, M.A. A systematic review of built environment factors related to physical activity and obesity risk: Implications for smart growth urban planning. *Obes. Rev.* **2011**, *12*, e173–e182.
25. King, A.C.; Sallis, J.F.; Frank, L.D.; Saelens, B.E.; Cain, K.; Conway, T.L.; Chapman, J.E.; Ahn, D.K.; Kerr, J. Aging in neighborhoods differing in walkability and income: Associations with physical activity and obesity in older adults. *Soc. Sci. Med.* **2011**, *73*, 1525–1533.
26. Sallis, J.F.; Floyd, M.F.; Rodriguez, D.A.; Saelens, B.E. Role of built environments in physical activity, obesity, and cardiovascular disease. *Circulation* **2012**, *125*, 729–737.
27. Knuiman, M.W.; Christian, H.E.; Divitini, M.L.; Foster, S.A.; Bull, F.C.; Badland, H.M.; Giles-Corti, B. A Longitudinal Analysis of the Influence of the Neighborhood Built Environment on Walking for TransportationThe RESIDE Study. *Am. J. Epidemiol.* **2014**, *180*, 453–461.
28. Mackenbach, J.D.; Rutter, H.; Compernolle, S.; Glonti, K.; Oppert, J.M.; Charreire, H.; De Bourdeaudhuij, I.; Brug, J.; Nijpels, G.; Lakerveld, J. Obesogenic environments: A systematic review of the association between the physical environment and adult weight status, the SPOTLIGHT project. *BMC Public Health* **2014**, *14*, 233.
29. Van Dyck, D.; Cerin, E.; De Bourdeaudhuij, I.; Hinckson, E.; Reis, R.S.; Davey, R.; Sarmiento, O.L.; Mitas, J.; Troelsen, J.; MacFarlane, D.; et al. International study of objectively measured physical activity and sedentary time with body mass index and obesity: IPEN adult study. *Int. J. Obes. (Lond.)* **2015**, *39*, 199–207.
30. Mayne, D.J.; Morgan, G.G.; Jalaludin, B.B.; Bauman, A.E. The contribution of area-level walkability to geographic variation in physical activity: A spatial analysis of 95,837 participants from the 45 and Up Study living in Sydney, Australia. *Popul. Health Metr.* **2017**, *15*, 38.
31. Merom, D.; Ding, D.; Corpuz, G.; Bauman, A. Walking in Sydney: Trends in prevalence by geographic areas using information from transport and health surveillance systems. *J. Transp. Health* **2015**, *2*, 350–359.
32. Sato, M.; Du, J.; Inoue, Y. Rate of Physical Activity and Community Health: Evidence from U.S. Counties. *J. Phys. Act. Health* **2016**, *13*, 640–648.
33. Dohrenwend, B.P.; Shrout, P.E.; Egri, G.; Mendelsohn, F.S. Nonspecific psychological distress and other dimensions of psychopathology: Measures for use in the general population. *Arch. Gen. Psychiatry* **1980**, *37*, 1229–1236.
34. Drapeau, A.; Marchand, A.; Beaulieu-Prévost, D. Epidemiology of psychological distress. In *Mental Illnesses—Understanding, Prediction and Control*; Book Section 5; L'Abate, L., Ed.; InTech: London, UK, 2012; pp. 105–134.
35. Centers for Disease Control and Prevention. Behavioral Risk Factor Surveillance System. Available online: https://www.cdc.gov/brfss/ (accessed on 31 January 2018).
36. Australian Bureau of Statistics. Australian Health Survey. Available online: http://www.abs.gov.au/australianhealthsurvey (accessed on 31 January 2018).
37. Pattenden, S.; Casson, K.; Cook, S.; Dolk, H. Geographical variation in infant mortality, stillbirth and low birth weight in Northern Ireland, 1992–2002. *J. Epidemiol. Community Health* **2011**, *65*, 1159–1165.
38. Bonney, A.; Mayne, D.J.; Jones, B.D.; Bott, L.; Andersen, S.E.; Caputi, P.; Weston, K.M.; Iverson, D.C. Area-Level Socioeconomic Gradients in Overweight and Obesity in a Community-Derived Cohort of Health Service Users—A Cross-Sectional Study. *PLoS ONE* **2015**, *10*, e0137261.
39. Beale, L.; Abellan, J.J.; Hodgson, S.; Jarup, L. Methodologic issues and approaches to spatial epidemiology. *Environ. Health Perspect.* **2008**, *116*, 1105–1110.

40. Chaix, B.; Leyland, A.H.; Sabel, C.E.; Chauvin, P.; Råstam, L.; Kristersson, H.; Merlo, J. Spatial clustering of mental disorders and associated characteristics of the neighbourhood context in Malmö, Sweden, in 2001. *J. Epidemiol. Community Health* **2006**, *60*, 427–435.

41. Cheung, Y.T.D.; Spittal, M.J.; Pirkis, J.; Yip, P.S.F. Spatial analysis of suicide mortality in Australia: Investigation of metropolitan-rural-remote differentials of suicide risk across states/territories. *Soc. Sci. Med.* **2012**, *75*, 1460–1468.

42. Ngamini Ngui, A.; Apparicio, P.; Moltchanova, E.; Vasiliadis, H.M. Spatial analysis of suicide mortality in Québec: Spatial clustering and area factor correlates. *Psychiatry Res.* **2014**, *220*, 20–30.

43. Gruebner, O.; Lowe, S.R.; Sampson, L.; Galea, S. The geography of post-disaster mental health: Spatial patterning of psychological vulnerability and resilience factors in New York City after Hurricane Sandy. *Int. J. Health Geogr.* **2015**, *14*, 16.

44. Australian Bureau of Statistics. *Statistical Geography: Volume 1—Australian Standard Geographical Classification (ASGC), July 2006 (Catalgue No. 1216.0)*; Commonwealth of Australia: Canberra, Australia, 2006.

45. Australian Bureau of Statistics. TableBuilder Basic. Available online: http://www.abs.gov.au/websitedbs/censushome.nsf/home/tablebuilder (accessed on 31 January 2018).

46. Australian Bureau of Statistics. *Statistical Geography: Volume 2—Census Geographic Areas, 2006 (Catalogue No. 2905.0)*; Commonwealth of Australia: Canberra, Australia, 2006.

47. James, P.; Berrigan, D.; Hart, J.E.; Aaron Hipp, J.; Hoehner, C.M.; Kerr, J.; Major, J.M.; Oka, M.; Laden, F. Effects of buffer size and shape on associations between the built environment and energy balance. *Health Place* **2014**, *27*, 162–170.

48. Villanueva, K.; Knuiman, M.; Nathan, A.; Giles-Corti, B.; Christian, H.; Foster, S.; Bull, F. The impact of neighborhood walkability on walking: Does it differ across adult life stage and does neighborhood buffer size matter? *Health Place* **2014**, *25*, 43–46.

49. 45 and Up Study Collaborators. Cohort profile: The 45 and Up Study. *Int. J. Epidemiol.* **2008**, *37*, 941–947.

50. 45 and Up Study. Researcher Toolkit. Available online: https://www.saxinstitute.org.au/our-work/45-up-study/for-researchers/ (accessed on 31 January 2018).

51. The 45 and Up Study. The 45 and Up Study Data Book—December 2011 Release. Available online: https://www.saxinstitute.org.au/our-work/45-up-study/data-book/ (accessed on 31 January 2018).

52. The 45 and Up Study. The 45 and Up Study Data Book—April 2010 Release. Available online: https://www.saxinstitute.org.au/our-work/45-up-study/data-book/ (accessed on 31 January 2018).

53. Australian Bureau of Statistics. *Socio-Economic Indexes for Areas (SEIFA)—Technical Paper, 2006*; Australian Bureau of Statistics: Canberra, Australia, 2008.

54. Kessler, R.C.; Andrews, G.; Colpe, L.J.; Hiripi, E.; Mroczek, D.K.; Normand, S.L.T.; Walters, E.E.; Zaslavsky, A.M. Short screening scales to monitor population prevalences and trends in non-specific psychological distress. *Psychol. Med.* **2002**, *32*, 959–976.

55. Australian Bureau of Statistics. Use of the Kessler Psychological Distress Scale in ABS Health Surveys, Australia, 2007–08 (Catalgue No. 4817.0.55.001). Available online: http://www.abs.gov.au/ausstats/abs@.nsf/mf/4817.0.55.001 (accessed on 31 January 2018).

56. Andrews, G.; Slade, T. Interpreting scores on the Kessler Psychological Distress Scale (K10). *Aust. N. Z. J. Public Health* **2001**, *25*, 494–497.

57. Centre for Epidemiology and Evidence. New South Wales Adult Population Health Survey. Available online: http://www.health.nsw.gov.au/surveys/adult/Pages/default.aspx (accessed on 31 January 2018).

58. Australian Bureau of Statistics. *National Survey of Mental Health and Wellbeing: Summary of Results*; Commonwealth of Australia: Canberra, Australia, 2007.

59. Australian Bureau of Statistics. Australian Health Survey: Users' Guide, 2011–13 (CATALOGUE NO. 4363.0.55.001). Available online: http://www.abs.gov.au/ausstats/abs@.nsf/mf/4363.0.55.001 (accessed on 31 January 2018).

60. Wooden, M. *Use of the Kessler Psychological Distress Scale in the HILDA Survey*; HILDA Project Discussion Paper Series No. 209; Melbourne Institute of Applied and Social Research, The University of Melbourne: Parkville, Australia, 2009.

61. Atkins, J.; Naismith, S.L.; Luscombe, G.M.; Hickie, I.B. Psychological distress and quality of life in older persons: Relative contributions of fixed and modifiable risk factors. *BMC Psychiatry* **2013**, *13*, 249.

62. Banks, E.; Byles, J.E.; Gibson, R.E.; Rodgers, B.; Latz, I.K.; Robinson, I.A.; Williamson, A.B.; Jorm, L.R. Is psychological distress in people living with cancer related to the fact of diagnosis, current treatment or level of disability? Findings from a large Australian study. *Med. J. Aust.* **2010**, *193*, S62–S67.

63. Banks, E.; Brown, A.; Attia, J.; Joshy, G.; Korda, R.; Reddy, P.; Paige, E. O168 Prospective investigation of psychological distress and incident cardiovascular disease hospitalisation and all-cause mortality, accounting for baseline physical impairment in 203,500 participants in the 45 and Up Study. *Glob. Heart* **2014**, *9*, e47–e48.

64. Byles, J.E.; Gallienne, L.; Blyth, F.M.; Banks, E. Relationship of age and gender to the prevalence and correlates of psychological distress in later life. *Int. Psychogeriatr.* **2012**, *24*, 1009–1018.

65. Byles, J.E.; Robinson, I.; Banks, E.; Gibson, R.; Leigh, L.; Rodgers, B.; Curryer, C.; Jorm, L. Psychological distress and comorbid physical conditions: Disease or disability? *Depress. Anxiety* **2014**, *31*, 524–532.

66. Feng, X.; Astell-Burt, T.; Kolt, G.S. Do social interactions explain ethnic differences in psychological distress and the protective effect of local ethnic density? A cross-sectional study of 226 487 adults in Australia. *BMJ Open* **2013**, *3*, e002713.

67. Feng, X.; Astell-Burt, T. What types of social interactions reduce the risk of psychological distress? Fixed effects longitudinal analysis of a cohort of 30,271 middle-to-older aged Australians. *J. Affect. Disord.* **2016**, *204*, 99–102.

68. George, E.S.; Jorm, L.; Kolt, G.S.; Bambrick, H.; Lujic, S. Physical activity and psychological distress in older men: Findings from the New South Wales 45 and up study. *J. Aging Phys. Act.* **2012**, *20*, 300–316.

69. Korda, R.J.; Paige, E.; Yiengprugsawan, V.; Latz, I.; Friel, S. Income-related inequalities in chronic conditions, physical functioning and psychological distress among older people in Australia: Cross-sectional findings from the 45 and up study. *BMC Public Health* **2014**, *14*, 1–10.

70. McNamara, B.J.; Banks, E.; Gubhaju, L.; Williamson, A.; Joshy, G.; Raphael, B.; Eades, S.J. Measuring psychological distress in older Aboriginal and Torres Strait Islanders Australians: A comparison of the K-10 and K-5. *Aust. N. Z. J. Public Health* **2014**, *38*, 567–573.

71. Paradise, M.B.; Glozier, N.S.; Naismith, S.L.; Davenport, T.A.; Hickie, I.B. Subjective memory complaints, vascular risk factors and psychological distress in the middle-aged: A cross-sectional study. *BMC Psychiatry* **2011**, *11*, 108.

72. Phongsavan, P.; Grunseit, A.C.; Bauman, A.; Broom, D.; Byles, J.; Clarke, J.; Redman, S.; Nutbeam, D. Age, Gender, Social Contacts, and Psychological Distress. *J. Aging Health* **2013**, *25*, 921–943.

73. Plotnikoff, R.C.; Costigan, S.A.; Short, C.; Grunseit, A.; James, E.; Johnson, N.; Bauman, A.; D'Este, C.; van der Ploeg, H.P.; Rhodes, R.E. Factors Associated with Higher Sitting Time in General, Chronic Disease, and Psychologically-Distressed, Adult Populations: Findings from the 45 & Up Study. *PLoS ONE* **2015**, *10*, e0127689.

74. Ussher, J.M.; Perz, J.; Kellett, A.; Chambers, S.; Latini, D.; Davis, I.D.; Rose, D.; Dowsett, G.W.; Williams, S. Health-Related Quality of Life, Psychological Distress, and Sexual Changes Following Prostate Cancer: A Comparison of Gay and Bisexual Men With Heterosexual Men. *J. Sex. Med.* **2016**, *13*, 425–434.

75. Ware, J.E., Jr.; Sherbourne, C.D. The MOS 36-item short-form health survey (SF-36). I. Conceptual framework and item selection. *Med. Care* **1992**, *30*, 473–483.

76. Ware, J.E.; Snow, K.K.; Kosinski, M.; Gandek, B. *SF-36 Health Survey: Manual and Interpretation Guide*; The Health Institute, New England Medical Center: Boston, MA, USA, 1993.

77. Leroux, B.G.; Lei, X.; Breslow, N. Estimation of Disease Rates in Small Areas: A new Mixed Model for Spatial Dependence. In *Statistical Models in Epidemiology, the Environment, and Clinical Trials*; Halloran, M.E., Berry, D., Eds.; Springer: New York, NY, USA, 2000; pp. 179–191.

78. Klassen, A.C.; Kulldorff, M.; Curriero, F. Geographical clustering of prostate cancer grade and stage at diagnosis, before and after adjustment for risk factors. *Int. J. Health Geogr.* **2005**, *4*, 1.

79. Waldhoer, T.; Wald, M.; Heinzl, H. Analysis of the spatial distribution of infant mortality by cause of death in Austria in 1984 to 2006. *Int. J. Health Geogr.* **2008**, *7*, 21.

80. Besag, J.; York, J.; Mollié, A. Bayesian image restoration, with two applications in spatial statistics. *Ann. Inst. Stat. Math.* **1991**, *43*, 1–20.

81. Lawson, A.; Browne, W.J.; Vidal Rodeiro, C.L. *Disease mapping with WinBUGS and MLwiN*; Statistics in Practice; Wiley: Hoboken, NJ, USA, 2003.

82. Waller, L.; Carlin, B. Disease mapping. In *Handbook of Spatial Statistics*; Chapman & Hall/CRC Handbooks of Modern Statistical Methods; Gelfand, A.E., Diggle, P.J., Feuentes, M., Guttorp, P., Eds.; CRC Press: Boca Raton, FL, USA, 2010; pp. 217–244.

83. Lunn, D.; Jackson, C.; Best, N.; Thomas, A.; Spiegelhalter, D. *The BUGS Book: A Practical Introduction to Bayesian Analysis*; Texts in Statistical Science; CRC Press: Boca Raton, FL, USA, 2012.

84. Gelman, A.; Rubin, D.B. Inference from iterative simulation using multiple sequences. *Stat. Sci.* **1992**, *7*, 457–511.

85. Cowles, M.K.; Carlin, B.P. Markov Chain Monte Carlo Convergence Diagnostics: A Comparative Review. *J. Am. Stat. Assoc.* **1996**, *91*, 883–904.

86. Mealing, N.M.; Banks, E.; Jorm, L.R.; Steel, D.G.; Clements, M.S.; Rogers, K.D. Investigation of relative risk estimates from studies of the same population with contrasting response rates and designs. *BMC Med. Res. Methodol.* **2010**, *10*, 1–12.

87. Spiegelhalter, D.J.; Best, N.G.; Carlin, B.P.; Van Der Linde, A. Bayesian measures of model complexity and fit. *J. R. Stat. Soc. Ser. B Stat. Methodol.* **2002**, *64*, 583–639.

88. Cramb, S.M.; Mengersen, K.L.; Baade, P.D. Developing the atlas of cancer in Queensland: Methodological issues. *Int. J. Health Geogr.* **2011**, *10*, 9.

89. Holowaty, E.J.; Norwood, T.A.; Wanigaratne, S.; Abellan, J.J.; Beale, L. Feasibility and utility of mapping disease risk at the neighbourhood level within a Canadian public health unit: An ecological study. *Int. J. Health Geogr.* **2010**, *9*, 21.

90. Chen, H.; Cohen, P.; Chen, S. How Big is a Big Odds Ratio? Interpreting the Magnitudes of Odds Ratios in Epidemiological Studies. *Commun. Stat. Simul. Comput.* **2010**, *39*, 860–864.

91. Sallis, J.F.; Bauman, A.; Pratt, M. Environmental and policy interventions to promote physical activity. *Am. J. Prev. Med.* **1998**, *15*, 379–397.

92. Ewing, R.; Cervero, R. Travel and the built environment: A synthesis. *Transp. Res. Rec.* **2001**, *1780*, 87–114.

93. Bauman, A.E.; Sallis, J.F.; Owen, N. Environmental and policy measurement in physical research. In *Physical Activity Assessments for Health-Related Research*; Welk, G., Ed.; Human Kinetics: Champaign, IL, USA, 2002; pp. 241–251.

94. Badland, H.; Schofield, G. Transport, urban design, and physical activity: An evidence-based update. *Transp. Res. Transp. Environ.* **2005**, *10*, 177–196.

95. Giles-Corti, B.; Macaulay, G.; Middleton, N.; Boruff, B.; Bull, F.; Butterworth, I.; Badland, H.; Mavoa, S.; Roberts, R.; Christian, H. Developing a research and practice tool to measure walkability: A demonstration project. *Health Promot. J. Austr.* **2014**, *25*, 160–166.

96. Riley, M.W. Special problems of sociological analysis. In *Sociological Research: A Case Approach*; Riley, M.W., Merton, R.K., Eds.; Harcourt, Brace, and World: New York, NY, USA, 1963; Volume 1, pp. 700–725.

97. Alker, H.A. A typology of ecological fallacies. In *Quantitative Ecological Analysis*; Dogan, M., Ed.; Massachusetts Institute of Technology: Cambridge, MA, USA, 1969; pp. 69–86.

98. Ding, D.; Gebel, K. Built environment, physical activity, and obesity: What have we learned from reviewing the literature? *Health Place* **2012**, *18*, 100–105.

99. Stockton, J.C.; Duke-Williams, O.; Stamatakis, E.; Mindell, J.S.; Brunner, E.J.; Shelton, N.J. Development of a novel walkability index for London, United Kingdom: cross-sectional application to the Whitehall II Study. *BMC Public Health* **2016**, *16*, 1–12.

100. Maxwell, S.E.; Cole, D.A. Bias in cross-sectional analyses of longitudinal mediation. *Psychol. Methods* **2007**, *12*, 23–44.

101. Lorant, V.; Deliège, D.; Eaton, W.; Robert, A.; Philippot, P.; Ansseau, M. Socioeconomic Inequalities in Depression: A Meta-Analysis. *Am. J. Epidemiol.* **2003**, *157*, 98–112.

102. Muntaner, C.; Eaton, W.W.; Miech, R.; O'Campo, P. Socioeconomic Position and Major Mental Disorders. *Epidemiol. Rev.* **2004**, *26*, 53–62.

103. Pickett, K.E.; Wilkinson, R.G. Inequality: An underacknowledged source of mental illness and distress. *Br. J. Psychiatry* **2010**, *197*, 426–428.

104. Campion, J.; Bhugra, D.; Bailey, S.; Marmot, M. Inequality and mental disorders: Opportunities for action. *Lancet* **2013**, *382*, 183–184.

105. World Health Organization. *Social Determinants of Mental Health*; World Health Organization: Geneva, Switzerland, 2014.

106. Silva, M.; Loureiro, A.; Cardoso, G. Social determinants of mental health: A review of the evidence. *Eur. J. Psychiatry* **2016**, *30*, 259–292.

107. Rai, D.; Zitko, P.; Jones, K.; Lynch, J.; Araya, R. Country- and individual-level socioeconomic determinants of depression: Multilevel cross-national comparison. *Br. J. Psychiatry* **2013**, *202*, 195–203.

108. Lee, E.W.; Park, J.H. Individual and Socioeconomic Contextual Effects on Depressive Symptom in Korea: Multilevel Analysis of Cross-sectional Nationwide Survey. *J. Korean Med. Sci.* **2015**, *30*, 186–193.

109. Gunn, J.M.; Ayton, D.R.; Densley, K.; Pallant, J.F.; Chondros, P.; Herrman, H.E.; Dowrick, C.F. The association between chronic illness, multimorbidity and depressive symptoms in an Australian primary care cohort. *Soc. Psychiatry Psychiatr. Epidemiol.* **2012**, *47*, 175–184.

110. Fortin, M.; Bravo, G.; Hudon, C.; Lapointe, L.; Dubois, M.F.; Almirall, J. Psychological Distress and Multimorbidity in Primary Care. *Ann. Fam. Med.* **2006**, *4*, 417–422.

111. Ormel, J.; Rijsdijk, F.V.; Sullivan, M.; van Sonderen, E.; Kempen, G.I.J.M. Temporal and Reciprocal Relationship Between IADL/ADL Disability and Depressive Symptoms in Late Life. *J. Gerontol. B Psychol. Sci. Soc. Sci.* **2002**, *57*, P338–P347.

112. Bazemore, A.; Phillips, R.L.; Miyoshi, T. Harnessing Geographic Information Systems (GIS) to Enable Community-Oriented Primary Care. *J. Am. Board Fam. Med.* **2010**, *23*, 22–31.

113. Criqui, M.H. Response bias and risk ratios in epidemiologic studies. *Am. J. Epidemiol.* **1979**, *109*, 394–399.

114. Nohr, E.A.; Frydenberg, M.; Henriksen, T.B.; Olsen, J. Does low participation in cohort studies induce bias? *Epidemiology* **2006**, *17*, 413–418.

115. Mayne, D.J.; Morgan, G.G.; Jalaludin, B.B.; Bauman, A.E. Is it worth the weight? Adjusting physical activity ratio estimates for individual-level non-response is not required in area-level spatial analyses of the 45 and Up Study cohort. Presented at the 45 and Up Study Annual Forum, Sydney, Australia, 24 October 2017.

116. Schwartz, S. The fallacy of the ecological fallacy: The potential misuse of a concept and the consequences. *Am. J. Public Health* **1994**, *84*, 819–824.

117. Langford, I.H.; Leyland, A.H.; Rasbash, J.; Goldstein, H. Multilevel Modelling of the Geographical Distributions of Diseases. *J. R. Stat. Soc. Ser. C Appl. Stat.* **1999**, *48*, 253–268.

118. Altman, D.G. Statistics in medical journals: Some recent trends. *Stat. Med.* **2000**, *19*, 3275–3289.

119. Kristman, V.; Manno, M.; Côté, P. Loss to Follow-Up in Cohort Studies: How Much is Too Much? *Eur. J. Epidemiol.* **2004**, *19*, 751–760.

120. Baron, R.M.; Kenny, D.A. The moderator–mediator variable distinction in social psychological research: Conceptual, strategic, and statistical considerations. *J. Pers. Soc. Psychol.* **1986**, *51*, 1173–1182.

121. MacKinnon, D.P.; Krull, J.L.; Lockwood, C.M. Equivalence of the Mediation, Confounding and Suppression Effect. *Prev. Sci.* **2000**, *1*, 173–181.

122. Huque, M.H.; Anderson, C.; Walton, R.; Ryan, L. Individual level covariate adjusted conditional autoregressive (indiCAR) model for disease mapping. *Int. J. Health Geogr.* **2016**, *15*, 25.

123. Rue, H.; Martino, S.; Chopin, N. Approximate Bayesian inference for latent Gaussian models by using integrated nested Laplace approximations. *J. R. Stat. Soc. Ser. B Stat. Methodol.* **2009**, *71*, 319–392.

124. Bivand, R.; Gómez-Rubio, V.; Rue, H. Spatial Data Analysis with R-INLA with Some Extensions. *J. Stat. Softw.* **2015**, *63*, 31.

125. Openshaw, S. *the Modifiable Areal Unit Problem*; (CATMOG 38); Geo Books: Norwich, UK, 1984.

MDPI

St. Alban-Anlage 66

4052 Basel

Switzerland

Tel. +41 61 683 77 34

Fax +41 61 302 89 18

www.mdpi.com

International Journal of Environmental Research and Public Health Editorial Office

E-mail: ijerph@mdpi.com

www.mdpi.com/journal/ijerph